ARRL's Tech Q&A

Your Quick & Easy Path to a Technician Ham Radio License

Written by: Ward Silver, NØAX

Editorial Assistants:

Maty Weinberg, KB1EIB

Production Staff:

David Pingree, N1NAS, Senior Technical Illustrator
Jodi Morin, KA1JPA, Assistant Production Supervisor: Layout
Kathy Ford, Proofreader
Sue Fagan, Graphic Design Supervisor: Cover Design
Michelle Bloom, WB1ENT, Production Supervisor: Layout

Published By:

ARRL *The national association for AMATEUR RADIO*

225 Main Street
Newington, CT 06111-1494

This work is publication No. 223 of the Radio Amateur's Library,

Printed in USA

ISBN:0-87259-964-7
Fourth Edition

This book may be used for Technician license exams given beginning July 1, 2006. *QST* and ARRLWeb (www.arrl.org) will have news about any Rules changes.

CONTENTS

FOREWORD

Congratulations! You have taken the first step into the exciting world of Amateur Radio! You are about to join nearly 700,000 licensed Amateur Radio operators in the United States and nearly three million people around the world who call themselves "hams." You will soon be communicating with some of those hams — in your community, in your part of the state or around the world.

There are many important reasons why governments around the world allow Amateur Radio operators to use valuable radio frequencies for personal communications. As a licensed Amateur Radio operator, you will become part of a large group of trained communicators and electronics technicians. You will be an important emergency communications resource for your neighbors and fellow citizens. Who knows when you may find yourself in the situation of having the only communications link outside your neighborhood? Whether you are caught in a flood, earthquake or other natural disaster or answering a call from someone else, you can provide the knowledge and resources to help. Don't wait for a real emergency to learn emergency communications methods, though. Practice your skills by participating in the many public service opportunities in your area. You might even want to sign up for ARRL's on-line Emergency Communications Course.

No matter what other interests you may have, you will find ways to tie them to Amateur Radio.

- ***Computers?*** Hams have been using computers to enhance their enjoyment of their hobby since the 1970s. Computer-based data modes are among the fastest growing methods of Amateur Radio communication.
- ***Video?*** Using computer software and video cameras, hams exchange moving or still images by using Amateur Television (ATV) or Slow-Scan Television (SSTV).
- ***Electronics?*** There is no better way to learn about electronics than to prepare for a license exam or to build a useful piece of equipment for your ham station.
- ***Hiking, biking or other outdoor activities?*** Many hams love nothing more than packing a wire antenna and a small, lightweight radio, and heading for the great outdoors.

Whether across town or across the sea, hams are always looking for new friends. So wherever you may happen to be, you are probably near someone — perhaps a whole club — who would be glad to help you get started. If you need help contacting hams, instructors, Volunteer Examiners or clubs in your area, contact us here at ARRL Headquarters. We'll help you get in touch with someone near you. (See the contact information at the bottom of the next page.)

When you pass that exam and enter the exciting world of Amateur Radio, you'll find plenty of activity to keep you busy. You'll also find plenty of friendly folks who are anxious to help you get started. Amateur Radio has many interesting areas to explore. You may be interested in one particular aspect of the hobby now, but be willing to try something new occasionally. You'll discover a world of unlimited potential!

Most of the active radio amateurs in the United States are members of ARRL. The hams' own organization since 1914, ARRL is truly the national association for Amateur Radio. We provide training materials and other services, and represent our members nationally and internationally. *ARRL's Tech Q & A* is just one of the many ARRL publications for all levels and interests in Amateur Radio. You don't need a ham license to join. If you're interested in ham radio, we're interested in you. It's as simple as that! We have included an invitation for you to join ARRL at the back of this book.

If you have comments or suggestions about this book, please use the Feedback Form at the back of this book. We'd like to hear from you. Thanks, and good luck!

David Sumner, K1ZZ
Executive Vice President
Newington, Connecticut
March 2006

New Ham Desk
ARRL Headquarters
225 Main Street
Newington, CT
06111-1494
(860) 594-0200

Prospective new amateurs call:
800-32-NEW-HAM (800-326-3942)
You can also contact us via e-mail:
newham@arrl.org
or check out **ARRLWeb**:
www.arrl.org

When to Expect New Books

A Question Pool Committee (QPC) consisting of representatives from the various Volunteer Examiner Coordinators (VECs) prepares the license question pools. The QPC establishes a schedule for revising and implementing new Question Pools. The current Question Pool revision schedule is as follows:

Question Pool	Current Study Guides	Valid Through
Technician (Element 2)	*The ARRL Ham Radio License Manual* *ARRL's Tech Q & A,* 4th Edition	**June 30, 2009**
General (Element 3)	*The ARRL General Class License Manual* 5th Edition *ARRL's General Q & A,* 2nd Edition *ARRL's General Class Video Course,* 4th Edition	**June 30, 2007**
Amateur Extra (Element 4)	*The ARRL Extra Class License Manual,*8th Edition *ARRL's Extra Q & A,* 1st Edition	**To Be Announced**

As new Question Pools are released, ARRL will produce new study materials before the effective dates of the new Pools. Until then, the current Question Pools will remain in use, and current ARRL study materials, including this book, will help you prepare for your exams.

As the new Question Pool schedules are confirmed, the information will be published in *QST* and on ARRLWeb at **www.arrl.org**.

How to Use This Book

To earn a Technician Amateur Radio license, you must pass the Technician written exam, FCC Element 2. You will have to know some basic electronics theory and Amateur Radio operating practices and procedures. In addition, you will need to learn about some of the rules and regulations governing the Amateur Service, as contained in Part 97 of Title 47 of the Code of Federal Regulations—the Federal Communications Commission (FCC) Rules.

The Element 2 exam consists of 35 questions about Amateur Radio rules, theory and practice. A passing score is 74%, so you must answer 26 of the 35 questions correctly to pass. (Another way to look at this is that you can get as many as 9 questions wrong, and still pass the test.)

The questions and multiple-choice answers in this book are printed exactly as they were written by the Volunteer Examiner Coordinators' Question Pool Committee, and exactly as they will appear on your exam. (Be careful, though. The letter positions of the answer choices may be scrambled, so you can't simply memorize an answer letter for each question.) In this book, the letter of the correct answer is printed in **boldface type** just before the explanation. If you want to study without knowing the correct answer right away, simply cover the answer letters with your hand or a slip of paper.

As you read the explanations for many of the questions you will find words printed in **boldface type**. These words are important terms, and will help you identify the correct answer to the question. Words in *italics* are important words for understanding the general topic.

Introduction

The Technician License

Earning a Technician Amateur Radio license is a good way to begin enjoying ham radio. There is no Morse code exam for this license, and the Element 2 written exam is not difficult. There is no difficult math or electronics background required. You are sure to find the operating privileges available to a Technician licensee to be worth the time spent learning about Amateur Radio. After passing the exam, you will be able to operate on *every frequency* above 50 megahertz that is assigned to the Amateur Radio Service. With full operating privileges on those bands, you'll be ready to experience the excitement of Amateur Radio!

Perhaps your interest is in Amateur or "ham" radio's long history of providing emergency communications in time of need. Your experience with computers might lead you to explore the many digital modes of amateur communication and networking. If your eyes turn to the stars on a clear night, you might enjoy tracking the amateur satellites and using them to relay your signals to other amateurs around the world! Your whole family can enjoy Amateur Radio, taking part in outdoor activities like ARRL Field Day and mobile operating during a vacation.

Once you make the commitment to study and learn what it takes to pass the exam, you *will* accomplish your goal. Many people pass the exam on their first try, so if you study the material and are prepared, chances are good that you will soon have your license. It may take you more than one attempt to pass the Technician license exam, but that's okay. There is no limit to how many times you can take it. Many Volunteer Examiner Teams have several exam versions available, so you may even be able to try the exam again at the same exam session. Time and available exam versions may limit the number of times you can try the exam at a single exam session. If you don't pass after a couple of tries you will certainly benefit from more study of the question pools before you try again.

An Overview of Amateur Radio

Earning an Amateur Radio license, at whatever level, is a special achievement. The nearly 700,000 people in the US who call themselves Amateur Radio operators, or hams, are part of a global fraternity. Radio amateurs provide a voluntary, noncommercial, communication service. This is especially true during natural disasters or other emergencies when the normal lines of communication are out of service. In the aftermath of hurricanes Katrina, Rita, and Wilma in 2005, more than 2000 hams traveled to the stricken areas to establish communications links until normal systems were restored. Thousands more relayed information around the country. In every county and city, organized groups of amateur operators train and prepare to support their communities during disasters and emergencies of every type.

Hams have made many important contributions to the field of electronics and communications, and this tradition continues today. Amateur Radio experimentation is yet another reason many people become part of this self-disciplined group of trained operators, technicians and electronics experts—an asset to any country. Hams pursue their hobby purely for personal enrichment in technical and operating skills, without any type of payment except the personal satisfaction they feel from a job well done!

Radio signals do not know territorial boundaries, so hams have a unique ability to enhance international goodwill. Hams become ambassadors of their country every time they put their stations on the air.

Amateur Radio has been around since the early 1900s. Hams have always been at the forefront of technology. Today, hams relay signals through their own satellites, bounce signals off the moon, relay messages automatically through computerized radio networks and use any number of other "exotic" communications techniques. Amateurs talk from hand-held transceivers through mountaintop repeater stations that can relay their signals to other hams' cars or homes or through the Internet around the world. Hams send their own pictures by television, talk with other hams around the world by voice or, keeping alive a distinctive traditional skill, tap out messages in Morse code.

The US government, through the Federal Communications Commission (FCC), grants all US Amateur Radio licenses. Amateurs are expected to know more about their equipment and operating techniques because of the tremendous flexibility granted to the Amateur Service. Unlike other radio services, amateurs organize their own methods of communication, they are encouraged to build and repair their own, perform experiments with antennas and with radio propagation, and invent their own protocols and networks. The FCC licensing procedure ensures that amateurs have the necessary operating skill and electronics know-how to use that flexibility wisely and not interfere with other radio services.

Who Can Be a Ham?

The FCC doesn't care how old you are or whether you're a US citizen. If you pass the examination, the Commission will issue you an amateur license. Any person (except the agent of a foreign government) may take the exam and, if successful, receive an amateur license. It's important to understand that if a citizen of a foreign country receives an amateur license in this manner, he or she is a US Amateur Radio operator. (This should not be confused with a reciprocal permit for alien amateur licensee, which allows visitors from certain countries who hold valid amateur licenses in their homelands to operate their own stations in the US without having to take an FCC exam.)

License Structure

Anyone earning a new Amateur Radio license can earn one of three license classes — Technician, General and Amateur Extra. These vary in degree of knowledge required and frequency privileges granted. Higher-class licenses have more comprehensive examinations. In return for passing a more difficult exam you earn more frequency privileges (frequency space in the radio spectrum). The vast majority of beginners start with the most basic license, the Technician. (You can take as many of the license class exams as you want!)

At the time this book was written, Technician licensees who also learned the international Morse code and passed an exam to demonstrate their knowledge of code at 5 words per minute, called Element 1, gained some frequency privileges on four of the amateur high-frequency (HF) bands. This license was previously called the Technician Plus license, and many amateurs will refer to it by that name.

It is possible that the FCC will eliminate the Morse code exam for all Amateur Radio licenses, but until that happens, the exam is still required if you wish to obtain HF privileges, or upgrade to General.

Table 1 lists the amateur license classes you can earn, along with a brief description of their exam requirements and operating privileges. On those four HF bands you will experience the thrill of *working* (contacting) other Amateur Radio operators in just about any country in the world. There's nothing quite like making friends with other amateurs around the world.

Although there are also other amateur license classes, the FCC is no longer issuing new licenses for these classes. The Novice license was long considered

Table 1

Amateur Operator Licenses[1]

Note: At the time of this writing, the FCC was considering the possible elimination of Morse code testing (Element 1). Contact the ARRL or see the ARRL Web at **www.arrl.org** for updates.

Class	*Code Test*	*Written Exam*	*Privileges*
Technician	None (Element 2)[2]	Basic theory and regulations	All above 50 MHz
Technician With Morse code	5 WPM (Element 1)	Basic theory and regulations (Element 2)[2]	All "Novice" HF and all above 50 MHz
General	5 WPM (Element 1)	Basic theory and regulations; General theory and regulations Elements 2 and 3)	All except those reserved for Advanced and Amateur Extra
Amateur Extra	5 WPM (Element 1)	All lower exam elements, plus Amateur Extra theory (Elements 2, 3 and 4)	All

[1]A licensed radio amateur will be required to pass only those elements that are not included in the examination for the amateur license currently held.

[2]If you have a Technician license issued before March 21, 1987, you also have credit for Elements 1 and 3. You must be able to prove that your Technician license was issued before March 21, 1987 to receive this credit.

Table 2

US Amateur Bands

June 1, 2003 Novice, Advanced and Technician Plus Allocations

New Novice, Advanced and Technician Plus licenses are no longer being issued, but *existing* Novice, Technician Plus and Advanced class licenses are unchanged. Amateurs can continue to renew these licenses. Technicians who pass the 5 wpm Morse code exam *after* that date have Technician Plus privileges, although their license says Technician. They must retain the 5 wpm Certificate of Successful Completion of Examination (CSCE) as proof. The CSCE is valid indefinitely for operating authorization, but is valid only for 365 days for upgrade credit.

160 METERS

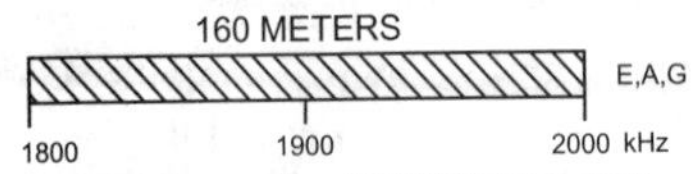

Amateur stations operating at 1900-2000 kHz must not cause harmful interference to the radiolocation service and are afforded no protection from radiolocation operations.

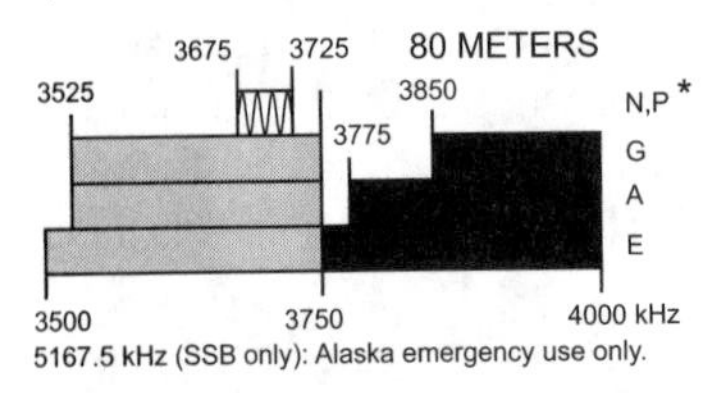

5167.5 kHz (SSB only): Alaska emergency use only.

60 METERS

General, Advanced, and Amateur Extra licensees may use the following five channels on a secondary basis with a maximum effective radiated power of 50 W PEP relative to a half wave dipole. Only upper sideband suppressed carrier voice transmissions may be used. The frequencies are 5330.5, 5346.5, 5366.5, 5371.5 and 5403.5 kHz. The occupied bandwidth is limited to 2.8 kHz centered on 5332, 5348, 5368, 5373, and 5405 kHz respectively.

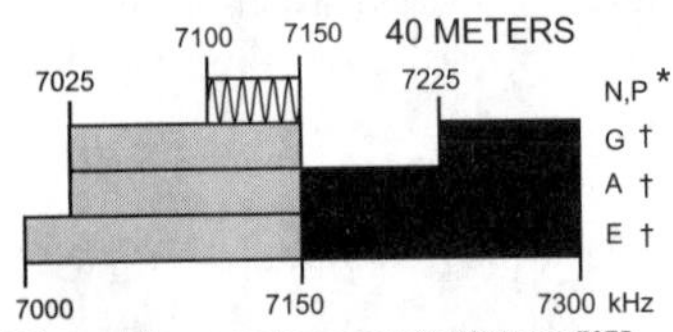

† Phone and Image modes are permitted between 7075 and 7100 kHz for FCC licensed stations in ITU Regions 1 and 3 and by FCC licensed stations in ITU Region 2 West of 130 degrees West longitude or South of 20 degrees North latitude. See Sections 97.305(c) and 97.307(f)(11). Novice and Technician Plus licensees outside ITU Region 2 may use CW only between 7050 and 7075 kHz. See Section 97.301(e). These exemptions do not apply to stations in the continental US.

30 METERS

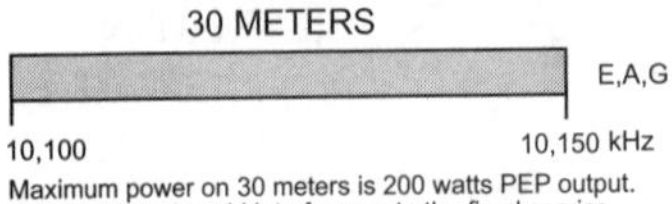

Maximum power on 30 meters is 200 watts PEP output. Amateurs must avoid interference to the fixed service outside the US.

20 METERS

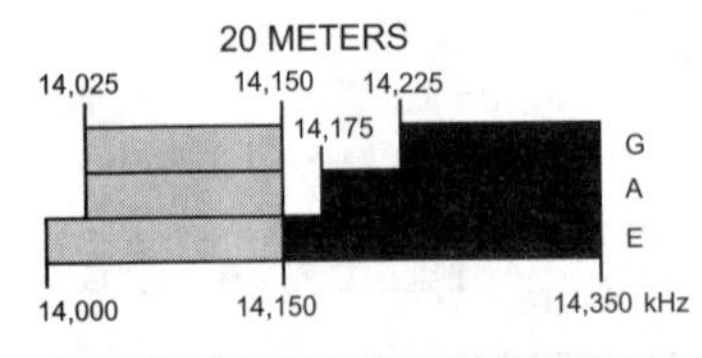

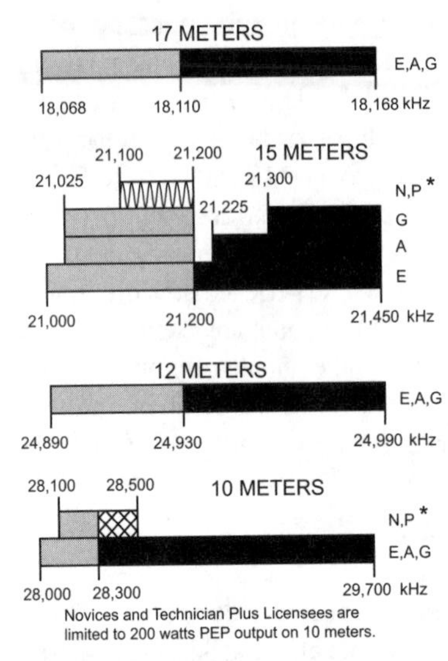

Novices and Technician Plus Licensees are limited to 200 watts PEP output on 10 meters.

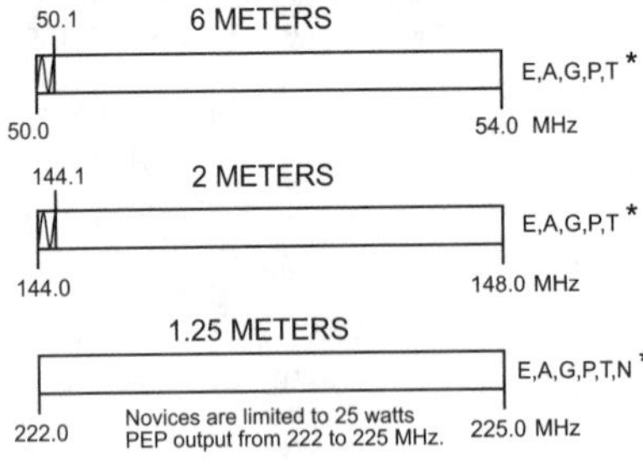

Novices are limited to 25 watts PEP output from 222 to 225 MHz.

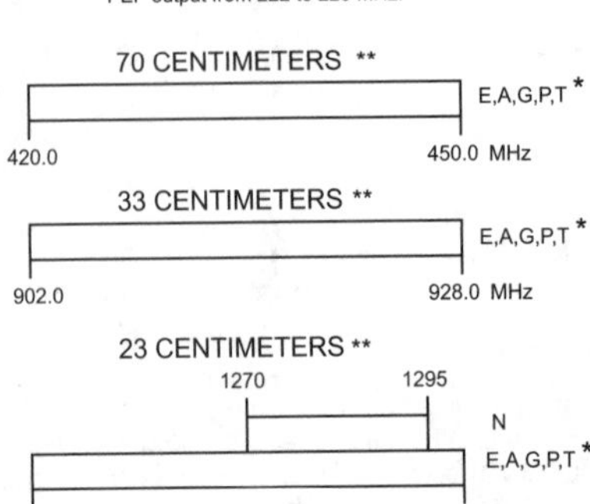

Novices are limited to 5 watts PEP output from 1270 to 1295 MHz.

ARRL *The national association for* **AMATEUR RADIO**

US AMATEUR POWER LIMITS

At all times, transmitter power should be kept down to that necessary to carry out the desired communications. Power is rated in watts PEP output. Unless otherwise stated, the maximum power output is 1500 W. Power for all license classes is limited to 200 W in the 10,100-10,150 kHz band and in all Novice subbands below 28,100 kHz. Novices and Technicians are restricted to 200 W in the 28,100-28,500 kHz subbands. In addition, Novices are restricted to 25 W in the 222-225 MHz band and 5 W in the 1270-1295 MHz subband.

KEY

= CW, RTTY and data

= CW, RTTY, data, MCW, test, phone and image

= CW, phone and image

= CW and SSB phone

= CW, RTTY, data, phone, and image

= CW only

E = EXTRA CLASS
A = ADVANCED
G = GENERAL
P = TECHNICIAN PLUS
T = TECHNICIAN
N = NOVICE

*Technicians who have passed the 5 wpm Morse code exam are indicated as "P".

**Geographical and power restrictions apply to all bands with frequencies above 420 MHz. See *The ARRL FCC Rule Book* for more information about your area.

***219-220 MHz allocated to amateurs on a secondary basis for fixed digital message forwarding systems only and can be operated by all licensees except Novices.

All licensees except Novices are authorized all modes on the following frequencies:

2300-2310 MHz
2390-2450 MHz
3300-3500 MHz
5650-5925 MHz
10.0-10.5 GHz
24.0-24.25 GHz
47.0-47.2 GHz
76-81.0 GHz
122.225-123.0 GHz
134-141 GHz
241-250 GHz
All above 275 GHz

the beginner's license. Exams for this license were discontinued as of April 15, 2000. The FCC also stopped issuing new Advanced class licenses on that date. They will continue to renew previously issued licenses, however, so you will probably meet some Novice and Advanced class licensees on the air.

The written Technician exam, called Element 2, covers some basic radio fundamentals and knowledge of some of the rules and regulations in Part 97 of the FCC Rules. With a little study you'll soon be ready to pass the Technician exam.

Each step up the Amateur Radio license ladder requires the applicant to pass the lower exams. So if you want to start out as a General class or even an Amateur Extra class licensee, you must also pass the Technician written exam. This does not mean you have to pass the Technician exam again if you already hold a Technician license! Your valid Amateur Radio license gives you credit for all the exam elements of that license when you go to upgrade. If you now hold a Technician license, you will only have to pass the Element 1 Morse code exam and the Element 3 General class written exam for a General class license.

A Technician license gives you the freedom to develop operating and technical skills through on-the-air experience. These skills will help you upgrade to a higher class of license and obtain additional privileges.

As a Technician, you can use a wide range of frequency bands — *all amateur bands above 50 MHz*, in fact. (See **Table 2** and **Figure 1**.) You'll be able to use point-to-point communications on VHF FM, communicate through repeaters, use packet radio, and access orbiting satellites to relay your signals over a wider area. You can use those skills to provide public service through emergency communications and message handling.

Learning Morse Code

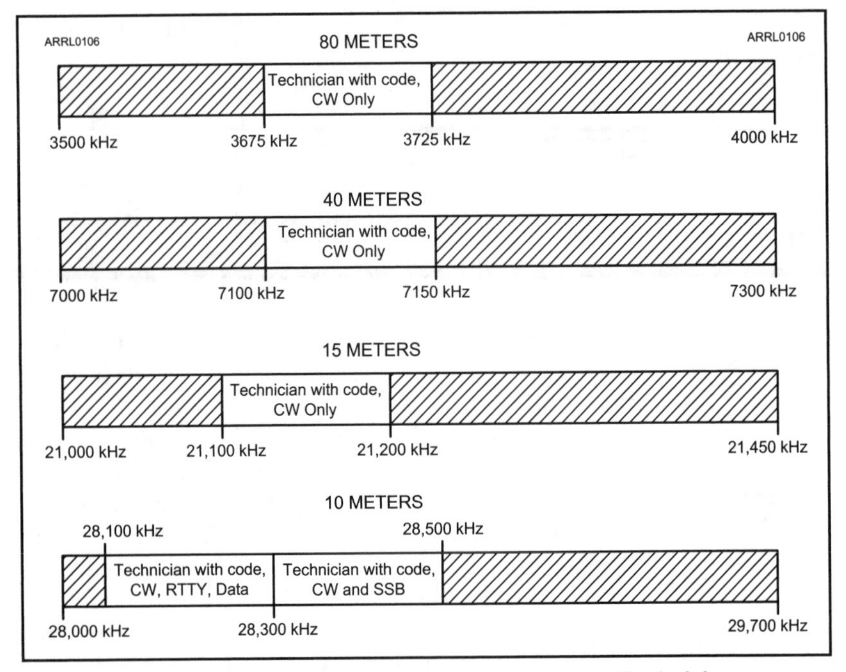

Figure 1—This chart details the HF privileges available to Technician licensees that pass their Element 1 Morse code test. When this book was written, however, the FCC was considering the possible elimination of Morse code testing. See the ARRL Web at www.arrl.org for updates.

Even if you don't plan to use Morse code now, there may come a time when you decide you would like to upgrade your license and earn some operating privileges on the high-frequency (HF) bands. Learning Morse code is a matter of practice. Instructions on learning the code, how to handle a telegraph key, and so on, can be found in *The ARRL General Class License Manual*, published by the ARRL. In addition, *Ham University* is a CD-based course that teaches you the Morse code letter-by-letter, by exercises, and even as a game. You will be ready to pass your 5 word-per-minute code exam when you finish the lessons in *Ham University*.

You can purchase any of these products from your local Amateur Radio equipment dealer or directly from the ARRL, 225 Main St, Newington, CT 06111. To place an order, call, toll-free, **888-277-5289**. You can also send e-mail to: **pubsales@arrl.org** or check out our World Wide Web site: **www.arrl.org**. Prospective new amateurs can call: **800-32-NEW HAM** (**800-326-3942**) for additional information.

Besides listening to code practice, some on-the-air operating experience will be a great help in building your code speed. When you are in the middle of a contact via Amateur Radio, and have to copy the code the other station is sending to continue the conversation, your copying ability will improve quickly! Many repeater stations identify themselves in Morse code, as well. Although you did not

Table 3

W1AW's schedule is at the same local time throughout the year. The schedule according to your local time will change if your local time does not have seasonal adjustments that are made at the same time as North American time changes between standard time and daylight time. From the first Sunday in April to the last Sunday in October, UTC = Eastern Time + 4 hours. For the rest of the year, UTC = Eastern Time + 5 hours.

W1AW Schedule

PACIFIC	MTN	CENT	EAST	MON	TUE	WED	THU	FRI
6 AM	7 AM	8 AM	9 AM		FAST CODE	SLOW CODE	FAST CODE	SLOW CODE
7 AM-1 PM	8 AM-2 PM	9 AM-3 PM	10 AM-4 PM	VISITING OPERATOR TIME (12 PM-1 PM CLOSED FOR LUNCH)				
1 PM	2 PM	3 PM	4 PM	FAST CODE	SLOW CODE	FAST CODE	SLOW CODE	FAST CODE
2 PM	3 PM	4 PM	5 PM	CODE BULLETIN				
3 PM	4 PM	5 PM	6 PM	DIGITAL BULLETIN				
4 PM	5 PM	6 PM	7 PM	SLOW CODE	FAST CODE	SLOW CODE	FAST CODE	SLOW CODE
5 PM	6 PM	7 PM	8 PM	CODE BULLETIN				
6 PM	7 PM	8 PM	9 PM	DIGITAL BULLETIN				
6:45 PM	7:45 PM	8:45 PM	9:45 PM	VOICE BULLETIN				
7 PM	8 PM	9 PM	10 PM	FAST CODE	SLOW CODE	FAST CODE	SLOW CODE	FAST CODE
8 PM	9 PM	10 PM	11 PM	CODE BULLETIN				

♦**Morse code transmissions:**

Frequencies are 1.8075, 3.5815, 7.0475, 14.0475, 18.0975, 21.0675, 28.0675 and 147.555 MHz.

Slow Code = practice sent at 5, 7 1/2, 10, 13 and 15 wpm.

Fast Code = practice sent at 35, 30, 25, 20, 15, 13 and 10 wpm.

Code practice text is from the pages of *QST*. The source is given at the beginning of each practice session and alternate speeds within each session. For example, "Text is from July 2001 *QST*, pages 9 and 81," indicates that the plain text is from the article on page 9 and mixed number/letter groups are from page 81.

Code bulletins are sent at 18 wpm.

W1AW qualifying runs are sent on the same frequencies as the Morse code transmissions. West Coast qualifying runs are transmitted on approximately 3.590 MHz by K6YR. See "Contest Corral" in this issue. At the beginning of each code practice session, the schedule for the next qualifying run is presented. Underline one minute of the highest speed you copied, certify that your copy was made without aid, and send it to ARRL for grading. Please include your name, call sign (if any) and complete mailing address. The fee structure is $10 for a certificate, and $7.50 for endorsements.

♦**Digital transmissions:**

Frequencies are 3.625, 7.095, 14.095, 18.1025, 21.095, 28.095 and 147.555 MHz.
Bulletins are sent at 45.45-baud Baudot and 100-baud AMTOR, FEC Mode B.
110-baud ASCII will be sent only as time allows.
On Tuesdays and Fridays at 6:30 PM Eastern Time, Keplerian elements for many amateur satellites are sent on the regular teleprinter frequencies.

♦ **Voice transmissions:**

Frequencies are 1.855, 3.99, 7.29, 14.29, 18.16, 21.39, 28.59 and 147.555 MHz.

♦ **Miscellanea:**

On Fridays, UTC, a DX bulletin replaces the regular bulletins.
W1AW is open to visitors 10 AM to noon and 1 PM to 3:45 PM on Monday through Friday. FCC licensed amateurs may operate the station during that time. Be sure to bring your current FCC amateur license or a photocopy. In a communication emergency, monitor W1AW for special bulletins as follows: voice on the hour, teleprinter at 15 minutes past the hour, and CW on the half hour.

Headquarters and W1AW are closed on New Year's Day, Presidents' Day, Good Friday, Memorial Day (May 30), Independence Day, Labor Day, Thanksgiving and the following Friday), and Christmas Day).

have to pass a Morse code test to earn your Technician license, there are no regulations prohibiting you from using code on the air. Many amateurs operate code on the VHF and UHF bands.

ARRL's Maxim Memorial Station, W1AW, transmits code practice and information bulletins of interest to all amateurs. These code-practice sessions and Morse code bulletins provide an excellent opportunity for code practice. **Table 3** is a W1AW operating schedule.

Station Call Signs

Many years ago, by international agreement, the nations of the world decided to allocate certain call sign prefixes to each country. This means that if you hear a radio station call sign beginning with W or K, for example, you know the station is licensed by the United States. A call sign beginning with the letter G is licensed by Great Britain, and a call sign beginning with VE is from Canada. *The ARRL DXCC List* is an operating aid no ham who is active on the HF bands should be without. That booklet, available from the ARRL, includes the common call-sign prefixes used by amateurs in virtually every location in the world. It also includes a check-off list to help you keep track of the countries you contact as you work toward collecting QSL cards from 100 or more countries to earn the prestigious DX Century Club award. (DX is ham lingo for distance, generally taken on the HF bands to mean any country outside the one from which you are operating.)

The International Telecommunication Union (ITU) radio regulations outline the basic principles used in forming amateur call signs. According to these regulations, an amateur call sign must be made up of one or two characters (the first one may be a numeral) as a prefix, followed by a numeral, and then a suffix of not more than three letters. The prefixes W, K, N and A are used in the United States. When the letter A is used in a US amateur call sign, it will always be with a two-letter prefix, AA to AL. The continental US is divided into 10 Amateur Radio call districts (sometimes called areas), numbered 0 through 9. **Figure 2** is a map showing the US call districts.

For information on the FCC's call-sign assignment system, and a table listing the blocks of call signs for each license class, see The ARRL *FCC Rule Book*. You may keep the same call sign when you change license class, if you wish. You must indicate that you want to receive a new call sign when you fill out an FCC Form 605 to apply for the exam or change your address.

The FCC also has a vanity call sign system. Under this system the FCC will issue a call sign selected from a list of preferred available call signs. While there is no fee for an Amateur Radio license, there is a fee for the selection of a vanity call sign. The current fee is $21.90 for a 10-year Amateur Radio license, paid upon application for a vanity call sign and at license renewal after that. (That fee may change as costs of administering the program change.) The latest details about the vanity call sign system are available from ARRL Regulatory Information, 225 Main Street, Newington, CT 06111-1494 and on ARRLWeb at **www.arrl.org**.

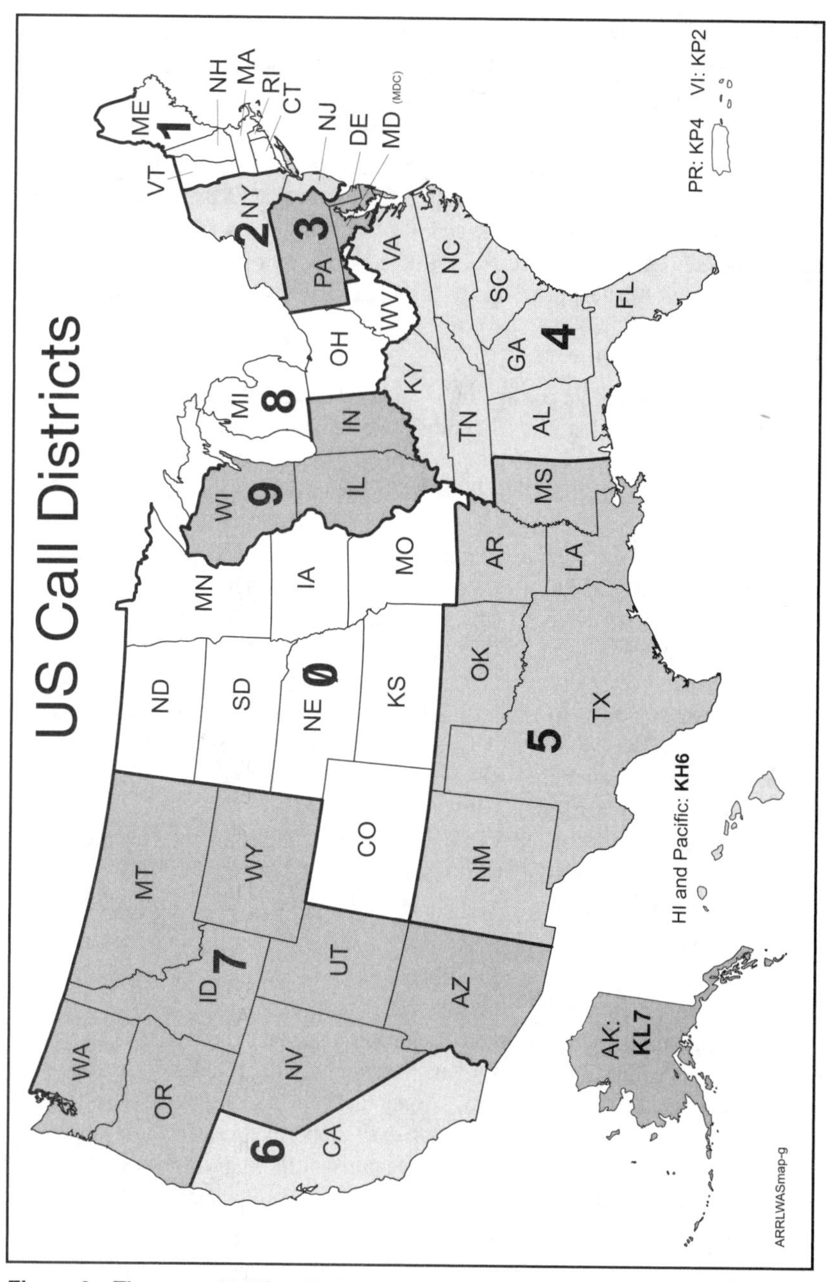

Figure 2—There are 10 US call sign areas. Hawaii is part of the sixth call area and Alaska is part of the seventh.

Earning a License

All US amateur exams are administered by Volunteer Examiners who are certified by a Volunteer-Examiner Coordinator (VEC). *The ARRL FCC Rule Book* contains more details about the Volunteer-Examiner program.

To qualify for a Technician license you must pass Element 2. If you already hold a valid Novice license, then you have credit for passing Element 1 and you can earn a Technician license with Morse code credit. In that case you will be able to continue using your Novice HF privileges. In addition, your upgrade to Technician will earn you full Amateur privileges on the VHF, UHF and higher frequency bands.

License Examinations

The Element 2 exam consists of 35 questions taken from a pool of more than 350 questions. A Question Pool Committee selected by the Volunteer Examiner Coordinators maintains the question pools for all amateur exams. The FCC allows Volunteer Examiners to select the questions for an amateur exam, but they must use the questions exactly as they are released by the VEC that coordinates the test session. If you attend a test session coordinated by the ARRL/VEC, your test will be designed by the ARRL/VEC or by a computer program designed by the VEC. The questions and answers will be exactly as they are printed in this book.

Before you can take an FCC exam, you'll have to fill out a copy of the National Conference of Volunteer Examiner Coordinators' (NCVEC) Quick Form 605. This form is used as an application for a new license or an upgraded license. The NCVEC Quick Form 605 is only used at license exam sessions. This form includes some information that the Volunteer Examiner Coordinator's office will need to process your application with the FCC. See **Figure 4**.

You should not use an NCVEC Quick Form 605 to apply for a license renewal or modification with the FCC. *Never* mail these forms to the FCC, because that will result in a rejection of the application. Likewise, an FCC Form 605 can't be used for an exam application.

Finding an Exam Opportunity

To determine where and when exams will be given in your area, contact the ARRL/VEC office, or watch for announcements in the Hamfest Calendar and Coming Conventions columns in *QST*. Many local clubs sponsor exams, so they are another good source of information on exam opportunities. Upcoming exams are listed on **ARRLWeb** at: **www.arrl.org/examsearch**. Registration deadlines, and the time and location of the exams, are mentioned prominently in publicity releases about upcoming sessions.

Taking the Exam

By the time examination day rolls around, you should have already prepared yourself. This means getting your schedule, supplies and mental attitude ready. Plan your schedule so you'll get to the examination site with plenty of time to spare. There's no harm in being early. In fact, you might have time to discuss hamming with another applicant, which is a great way to calm pretest nerves. Try not to discuss the material that will be on the examination, as this may make you even more nervous. By this time, it's too late to study anyway!

What supplies will you need? First, be sure you bring your current original Amateur Radio license, if you have one. Bring a photocopy of your license, too, as well as the original and a photocopy of any Certificates of Successful Completion of Examination (CSCE) that you plan to use for exam credit. Bring along several sharpened number 2 pencils and two pens (blue or black ink). Be sure to have a good eraser. A pocket calculator may also come in handy. You may use a programmable calculator if that is the kind you have, but take it into your exam "empty" (cleared of all programs and constants in memory). Don't program equations ahead of time, because you may be asked to demonstrate that there is nothing in the calculator memory. The examining team has the right to refuse a candidate the use of any calculator that they feel may contain information for the test or could otherwise be used to cheat on the exam.

The Volunteer Examiner Team is required to check two forms of identification before you enter the test room. This includes your *original* Amateur Radio license, if you have one—not a photocopy. A photo ID of some type is best for the second form of ID, but is not required by the FCC. Other acceptable forms of identification include a driver's license, a piece of mail addressed to you or a birth certificate.

The following description of the testing procedure applies to exams coordinated by the ARRL/VEC, although many other VECs use a similar procedure.

Code Test

If you need to take a Morse code test, you'll be handed a piece of paper to copy the code as it is sent. The test will begin with about a minute of practice copy. Then comes the actual test: at least five minutes of Morse code. You are responsible for knowing the 26 letters of the alphabet, the numerals 0 through 9, the period, comma, question mark, and the procedural signals $\overline{\text{AR}}$, $\overline{\text{SK}}$, $\overline{\text{BT}}$ (= or double dash) and $\overline{\text{DN}}$ (/ or fraction bar, sometimes called the "slant bar").

You may copy the entire text word for word, or just take notes on the content. At the end of the transmission, the examiner will hand you 10 questions about the text. Fill in the blanks with your answers. (You must spell each answer exactly as it was sent.) If you get at least 7 correct, you pass! Alternatively, the exam team has the option to look at your copy sheet if you fail the 10-question exam. If you have one minute of solid copy (25 characters), the examiners can certify that you passed the test on that basis. The format of the test transmission is similar to one side of a normal on-the-air amateur conversation.

A sending test may not be required. The Commission has decided that if applicants can demonstrate receiving ability, they most likely can also send at that speed. But be prepared for a sending test, just in case! Subpart 97.503(a) of the FCC Rules says, "A telegraphy examination must be sufficient to prove that the examinee has the ability to send correctly by hand and to receive correctly by ear

FCC 605
Main Form

Quick-Form Application for Authorization in the Ship, Aircraft, Amateur, Restricted and Commercial Operator, and General Mobile Radio Services

Approved by OMB
3060 - 0850
See instructions for public burden estimate

1) Radio Service Code: HA

Application Purpose (Select only one) (MD)

2) **NE** – New **MD** – Modification **AM** – Amendment **RO** – Renewal Only **RM** – Renewal / Modification **CA** – Cancellation of License **WD** – Withdrawal of Application **DU** – Duplicate License **AU** – Administrative Update

3) If this request if for Developmental License or STA (Special Temporary Authorization) enter the appropriate code and attach the required exhibit as described in the instructions. Otherwise enter 'N' (Not Applicable).	(N) D S N/A
4) If this request is for an Amendment or Withdrawal of Application, enter the file number of the pending application currently on file with the FCC.	File Number
5) If this request is for a Modification, Renewal Only, Renewal / Modification, Cancellation of License, Duplicate License, or Administrative Update, enter the call sign (serial number for Commercial Operator) of the existing FCC license. If this is a request for consolidation of DO & DM Operator Licenses, enter serial number of DO. Also, if filing for a ship exemption, you must provide call sign.	Call Sign/Serial # AB1FM
6) If this request is for a New, Amendment, Renewal Only, or Renewal Modification, enter the requested expiration date of the authorization (this item is optional).	MM DD
7) Does this filing request a Waiver of the Commission's rules? If 'Y', attach the required showing as described in the instructions.	(N) Yes No
8) Are attachments (other than associated schedules) being filed with this application?	(N) Yes No

Applicant/Licensee Information

9) FCC Registration Number (FRN): 0012345678

10) Applicant/Licensee legal entity type: (Select One)
☒ Individual ☐ Corporation ☐ Unincorporated Association ☐ Trust ☐ Government Entity
☐ Consortium ☐ General Partnership ☐ Limited Liability Company ☐ Limited Liability Partnership
☐ Limited Partnership ☐ Other (Description of Legal Entity) ____

11) First Name (if individual): MARIA MI: A Last Name: SOMMA Suffix:

12) Entity Name (if other than individual):

13) If the licensee name is being updated, is the update a result from the sale (or transfer of control) of the license(s) to another party and for which proper Commission approval has not been received or proper notification not provided? ( ) Yes No

14) Attention To:

15) P.O. Box: And/Or 16) Street Address: 225 MAIN ST.

17) City: NEWINGTON 18) State: CT 19) Zip Code/Postal Code: 06066 20) Country:

21) Telephone Number: 22) FAX Number:

23) E-Mail Address:

Ship Applicants/Licensees Only

24) Enter new name of vessel: ____

Aircraft Applicants/Licensees Only

25) Enter the new FAA Registration Number (the N-number): ____
NOTE: Do not enter the leading "N".

FCC 605 – Main Form
July 2005 - Page 1

Figure 3—Portions of FCC Form 605 showing the sections you would complete for a modification of your license, such as a change of address.

texts in the international Morse code at not less than the prescribed speed..."

Written Tests

The examiner will give each applicant a test booklet, an answer sheet and scratch paper. After that, you're on your own. The first thing to do is read the instructions. Be sure to sign your name every place it's called for. Do all of this at the beginning to get it out of the way.

Fee Status

26) Is the applicant/licensee exempt from FCC application Fees?	(N) Yes No
27) Is the applicant/licensee exempt from FCC regulatory Fees?	(N) Yes No

General Certification Statements

1) The applicant/licensee waives any claim to the use of any particular frequency or of the electromagnetic spectrum as against the regulatory power of the United States because of the previous use of the same, whether by license or otherwise, and requests an authorization in accordance with this application.
2) The applicant/licensee certifies that all statements made in this application and in the exhibits, attachments, or documents incorporated by reference are material, are part of this application, and are true, complete, correct, and made in good faith.
3) Neither the applicant/licensee nor any member thereof is a foreign government or a representative thereof.
4) The applicant/licensee certifies that neither the applicant/licensee nor any other party to the application is subject to a denial of Federal benefits pursuant to Section 5301 of the Anti-Drug Abuse Act of 1988, 21 U.S.C. § 862, because of a conviction for possession or distribution of a controlled substance. **This certification does not apply to applications filed in services exempted under Section 1.2002(c) of the rules, 47 CFR § 1.2002(c).** See Section 1.2002(b) of the rules, 47 CFR § 1.2002(b), for the definition of "party to the application" as used in this certification.
5) Amateur or GMRS applicant/licensee certifies that the construction of the station would NOT be an action which is likely to have a significant environmental effect (see the Commission's rules 47 CFR Sections 1.1301-1.1319 and Section 97.13(a) rules (available at web site http://wireless.fcc.gov/rules.html).
6) Amateur applicant/licensee certifies that they have READ and WILL COMPLY WITH Section 97.13(c) of the Commission's rules (available at web site http://wireless.fcc.gov/rules.html) regarding RADIOFREQUENCY (RF) RADIATION SAFETY and the amateur service section of OST/OET Bulletin Number 65 (available at web site http://www.fcc.gov/oet/info/documents/bulletins/).

Certification Statements For GMRS Applicants/Licensees

1) Applicant/Licensee certifies that he or she is claiming eligibility under Rule Section 95.5 of the Commission's rules.
2) Applicant/Licensee certifies that he or she is at least 18 years of age.
3) Applicant/Licensee certifies that he or she will comply with the requirement that use of frequencies 462.650, 467.650, 462.700 and 467.700 MHz is not permitted near the Canadian border North of Line A and East of Line C. These frequencies are used throughout Canada and harmful interference is anticipated.
4) Non-Individual applicants/licensees certify that they have NOT changed frequency or channel pairs, type of emission, antenna height, location of fixed transmitters, number of mobile units, area of mobile operation, or increase in power.

Certification Statements for Ship Applicants/Licensees (Including Ship Exemptions)

1) Applicant/Licensee certifies that they are the owner or operator of the vessel, a subsidiary communications corporation of the owner or operator of the vessel, a state or local government subdivision, or an agency of the US Government subject to Section 301 of the Communications Act.
2) This application is filed with the understanding that any action by the Commission thereon shall be limited to the voyage(s) described herein, and that apart from the provisions of the specific law from which the applicant/licensee requests an exemption, the vessel is in full compliance with all applicable statues, international agreements and regulations.

Signature

28) Typed or Printed Name of Party Authorized to Sign

First Name: MARIA	MI: A	Last Name: SOMMA	Suffix:
29) Title:			
Signature: Maria Somma			30) Date: 04/01/06

Failure to Sign This Application May Result in Dismissal Of The Application And Forfeiture Of Any Fees Paid

WILLFUL FALSE STATEMENTS MADE ON THIS FORM OR ANY ATTACHMENTS ARE PUNISHABLE BY FINE AND/OR IMPRISONMENT (U.S. Code, Title 18, Section 1001) AND / OR REVOCATION OF ANY STATION LICENSE OR CONSTRUCTION PERMIT (U.S. Code, Title 47, Section 312(a)(1)), AND / OR FORFEITURE (U.S. Code, Title 47, Section 503).

FCC 605 – Main Form
July 2005 - Page 2

Next, check the examination to see that all pages and questions are there. If not, report this to the examiner immediately. When filling in your answer sheet make sure your answers are marked next to the numbers that correspond to each question.

Go through the entire exam, and answer the easy questions first. Next, go back to the beginning and try the harder questions. Leave the really tough questions for last. Guessing can only help, as there is no additional penalty for answering incorrectly.

If you have to guess, do it intelligently: At first glance, you may find that you can eliminate one or more "distractors." Of the remaining responses, more than one may seem correct; only one is the best answer, however. To the applicant who is fully prepared, incorrect distracters to each question are obvious. Nothing beats preparation!

After you've finished, check the examination thoroughly. You may have read a question wrong or goofed in your arithmetic. Don't be overconfident. There's no rush, so take your time. Think, and check your answer sheet. When you feel you've done your best and can do no more, return the test booklet, answer sheet and scratch pad to the examiner.

The Volunteer-Examiner Team will grade the exam while you wait. The passing mark is 74%. (That means 26 out of 35 questions correct — or no more than 9 incorrect answers on the Element 2 exam.) You will receive a Certificate of Successful Completion of Examination (CSCE) showing all exam elements that you pass at that exam session.

If you are already licensed, and you pass the exam elements required to earn a higher license class, the CSCE authorizes you to operate with your new privileges immediately. When you use these new privileges, you must sign your call sign followed by the slant mark ("/"; on voice, say "stroke" or "slant") and the letters "KT," if you are upgrading from a Novice to a Technician with code license. You only have to follow this special identification procedure until the FCC grants your new license, however.

If you pass only some of the exam elements required for a license, you will still receive a CSCE. That certificate shows what exam elements you passed, and is valid for 365 days. Use it as proof that you passed those exam elements so you won't have to take them over again next time you try for the license.

Forms and Procedures

To renew or modify a license, you can file a copy of FCC Form 605. In addition, hams who have held a valid license that has expired within the past two years may apply for reinstatement with an FCC Form 605.

Licenses are normally good for ten years. Your application for a license renewal must be submitted to the FCC no more than 90 days before the license expires. (We recommend you submit the application for renewal between 90 and 60 days before your license expires.) If the FCC receives your renewal application before the license expires, you may continue to operate until your new license arrives, even if it is past the expiration date.

If you forget to apply before your license expires, you may still be able to renew your license without taking another exam. There is a two-year grace period, during which you may apply for renewal of your expired license. Use an FCC Form 605 to apply for reinstatement (and your old call sign). If you apply for reinstatement of your expired license under this two-year grace period, you may not operate your station until your new license is issued.

If you move or change addresses you should use an FCC Form 605 to notify the FCC of the change. If your license is lost or destroyed, however, just write a letter to the FCC explaining why you are requesting a new copy of your license.

You can ask one of the Volunteer Examiner Coordinators' offices to file your renewal application electronically if you don't want to mail the form to the FCC. You must still mail the form to the VEC, however. The ARRL/VEC Office will electronically file application forms. This service is free for any ARRL member.

Electronic Filing

You can also file your license renewal or address modification using the Universal Licensing System (ULS) on the World Wide Web. To use ULS, you must have an FCC Registration Number, or FRN. Obtain your FRN by registering with the Commission Registration System, known as CORES.

Described as an agency-wide registration system for anyone filing applications with or making payments to the FCC, CORES will assign a unique 10-digit FCC Registration Number (FRN) to all registrants. All Commission systems that handle financial, authorization of service, and enforcement activities will use the FRN. The FCC says use of the FRN will allow it to more rapidly verify fee payment. Amateurs mailing payments to the FCC — for example as part of a vanity call sign application — would include their FRN on FCC Form 159.

The on-line filing system and further information about CORES is available by visiting the FCC Web site, **www.fcc.gov** and clicking on the Commission Registration System link. Follow the directions on the Web site. It is also possible to register on CORES using a paper Form 160.

When you register with CORES you must supply a Taxpayer Identification Number, or TIN. For individuals, this is usually a Social Security Number. Club stations that do not have an EIN register as exempt.

Anyone can register on CORES and obtain an FRN. CORES/FRN is "entity registration." You don't need a license to be registered.

Once you have registered on CORES and obtained your FRN, you can proceed to renew or modify your license using the Universal Licensing System (ULS), also on the World Wide Web. Go to **www.fcc.gov/uls** and click on the "Online Filing" button. Follow the directions provided on the Web page to connect to the FCC's ULS database.

Paper Filing

The FCC has a set of detailed instructions for the Form 605, which are included with the form. To obtain a new Form 605, call the FCC Forms Distribution Center at 800-418-3676. You can also write to: Federal Communications Commission, Forms Distribution Center, 9300 E. Hampton Drive Capital Heights, MD 20743 (specify "Form 605" on the envelope). The Form 605 also is available from the FCC's fax on demand service. Call 202-418-0177 and ask for form number 000605. Form 605 also is available via the Internet. The World Wide Web location is: **www.fcc.gov/formpage.html** or you can receive the form via ftp to: **ftp.fcc.gov/pub/Forms/Form605**.

The ARRL/VEC has created a package that includes the portions of FCC

Form 605 that are needed for amateur applications, as well as a condensed set of instructions for completing the form. Write to: ARRL/VEC, Form 605, 225 Main Street, Newington, CT 06111-1494. (Please include a large business-sized stamped self-addressed envelope with your request.) **Figure 3** is a sample of those portions of an FCC Form 605 that you would complete to submit a change of address to the FCC.

Most of the form is simple to fill out. You will need to know that the Radio Service Code for box 1 is HA for Amateur Radio. (Just remember HAm radio.) Register with CORES as described above. When you receive your FRN from the FCC, you must include FRN on form.

The telephone number, fax number and e-mail address information is optional. The FCC will use that information to contact you in case there is a problem with your application.

Page two of the Form includes six General Certification Statements. Statement five may seem confusing. Basically, this statement means that you do not plan to install an antenna over 200 feet high, and that your permanent station location will not be in a designated wilderness area, wildlife preserve or nationally recognized scenic and recreational area.

The sixth statement indicates that you are familiar with the FCC RF Safety Rules, and that you will obey them. Chapter 10 (Subelement T0) of this book includes exam questions and explanations about the RF Safety Rules.

And Now, Let's Begin

The complete Technician question pool (Element 2) is printed in this book. Each chapter lists all the questions for a particular subelement (such as Control Operator Duties — T5). A brief explanation about the correct answer is given after each question.

Table 4 shows the study guide or syllabus for the Element 2 exam as released by the Volunteer-Examiner Coordinators' Question Pool Committee in December 2005. The syllabus lists the topics to be covered by the Technician exam, and so forms the basic outline for the remainder of this book. Use the syllabus to guide your study.

The question numbers used in the question pool refer to this syllabus. Each question number begins with a syllabus-point number (for example, T0C or T1A). The question numbers end with a two-digit number. For example, question T3B09 is the ninth question about the T3B syllabus point.

The Question Pool Committee designed the syllabus and question pool so there are the same number of points in each subelement as there are exam questions from that subelement. For example, two exam questions on the Technician exam must be from the "Radio Phenomena" subelement, so there are two groups for that point. These are numbered T3A and T3B. While not a requirement of the FCC Rules, the Question Pool Committee recommends that one question be taken from each group to make the best possible license exams.

Good luck with your studies!

NCVEC QUICK-FORM 605 APPLICATION FOR AMATEUR OPERATOR/PRIMARY STATION LICENSE

SECTION 1 - TO BE COMPLETED BY APPLICANT

PRINT LAST NAME / SUFFIX	FIRST NAME	INITIAL	STATION CALL SIGN (IF ANY)
Sayad	Daniel	S	

MAILING ADDRESS (Number and Street or P.O. Box)	SOCIAL SECURITY NUMBER / TIN (OR FCC LICENSEE ID #)
225 Main St	125-4236-54

CITY	STATE CODE	ZIP CODE (5 or 9 Numbers)	E-MAIL ADDRESS (OPTIONAL)
Newington	CT	06111	

DAYTIME TELEPHONE NUMBER (Include Area Code) OPTIONAL	FAX NUMBER (Include Area Code) OPTIONAL	ENTITY NAME (IF CLUB, MILITARY RECREATION, RACES)
860-594-0200		

Type of Applicant: [✓] Individual [ ] Amateur Club [ ] Military Recreation [ ] RACES (Modify Only)

CLUB, MILITARY RECREATION, OR RACES CALL SIGN

SIGNATURE OF RESPONSIBLE CLUB OFFICIAL

I HEREBY APPLY FOR (Make an X in the appropriate box(es))

- [✓] **EXAMINATION** for a **new** license grant
- [ ] **EXAMINATION** for **upgrade** of my license class
- [ ] **CHANGE** my **name** on my license to my new name

  Former Name: ____________ (Last name) (Suffix) (First name) (MI)
- [ ] **CHANGE** my mailing address to **above** address
- [ ] **CHANGE** my station **call sign** systematically

  Applicant's Initials: ____________
- [ ] **RENEWAL** of my license grant.

Do you have another license application on file with the FCC which has not been acted upon?	PURPOSE OF OTHER APPLICATION	PENDING FILE NUMBER (FOR VEC USE ONLY)

I certify that:

- I waive any claim to the use of any particular frequency regardless of prior use by license or otherwise;
- All statements and attachments are true, complete and correct to the best of my knowledge and belief and are made in good faith;
- I am not a representative of a foreign government;
- I am not subject to a denial of Federal benefits pursuant to Section 5301of the Anti-Drug Abuse Act of 1988, 21 U.S.C. § 862;
- The construction of my station will NOT be an action which is likely to have a significant environmental effect (See 47 CFR Sections 1.301-1.319 and Section 97.13(a));
- I have read and WILL COMPLY with Section 97.13(c) of the Commission's Rules regarding RADIOFREQUENCY (RF) RADIATION SAFETY and the amateur service section of OST/OET Bulletin Number 65.

Signature of applicant (Do not print, type, or stamp. Must match applicant's name above.)

X D Sayad Date Signed: 1/16/03

SECTION 2 - TO BE COMPLETED BY ALL ADMINISTERING VEs

Applicant is qualified for operator license class:

- [ ] **NO NEW LICENSE OR UPGRADE WAS EARNED**
- [X] **TECHNICIAN** Element 2
- [ ] **GENERAL** Elements 1, 2 and 3
- [ ] **AMATEUR EXTRA** Elements 1, 2, 3 and 4

DATE OF EXAMINATION SESSION: 1/16/03

EXAMINATION SESSION LOCATION: Newington, CT

VEC ORGANIZATION: ARRL

VEC RECEIPT DATE:

I CERTIFY THAT I HAVE COMPLIED WITH THE ADMINISTERING VE REQUIRMENTS IN PART 97 OF THE COMMISSION'S RULES AND WITH THE INSTRUCTIONS PROVIDED BY THE COORDINATING VEC AND THE FCC.

VEs NAME (Print First, MI, Last, Suffix)	VEs STATION CALL SIGN	VEs SIGNATURE (Must match name)	DATE SIGNED
1st: DAVID C PATTON	NT1N	David C Patton	16 JAN 03
2nd: PERRY T GREEN	WY1O	Perry T Green	1/16/03
3rd: Larry D. Wolfgang	WR1B	Larry D Wolfgang	1/16/03

DO NOT SEND THIS FORM TO FCC – THIS IS NOT AN FCC FORM.
IF THIS FORM IS SENT TO FCC, FCC WILL RETURN IT TO YOU WITHOUT ACTION.

NCVEC FORM 605 - FEBRUARY 2001
FOR VE/VEC USE ONLY - Page 1

Figure 4—An NCVEC Quick Form 605 as it would be completed for a new Technician license.

Table 4
Technician Class (Element 2) Syllabus

(Required for all operator licenses)

SUBELEMENT T1 – FCC Rules & Station License Responsibilities

[4 exam questions – 4 groups]

T1A - Basis and purpose of the Amateur Radio Service, penalties for unlicensed operation, other penalties, examinations

T1B - ITU regions, international regulations, US call sign structure, special event calls, vanity call signs

T1C – Authorized frequencies (Technician), reciprocal licensing, operation near band edges, spectrum sharing

T1D - The station license, correct name and address on file, license term, renewals, grace period

SUBELEMENT T2 - Control Operator Duties

[4 exam questions – 4 groups]

T2A - Prohibited communications: music, broadcasting, codes and ciphers, business use, permissible communications, bulletins, code practice, incidental music

T2B - Basic identification requirements, repeater ID standards, identification for non-voice modes, identification requirements for mobile and portable operation

T2C – Definition of control operator, location of control operator, automatic and remote control, auxiliary stations

T2D - Operating another person's station, guest operators at your station, third party communications, autopatch, incidental business use, compensation of operators, club stations, station security, station inspection, protection against unauthorized transmissions

SUBELEMENT T3 – Operating Practices

[4 exam questions – 4 groups]

T3A - Choosing an operating frequency, calling CQ, calling another station, test transmissions

T3B - Use of minimum power, band plans, repeater coordination, mode restricted sub-bands

T3C - Courtesy and respect for others, sensitive subject areas, obscene and indecent language

T3D - Interference to and from consumer devices, public relations, intentional and unintentional interference

SUBELEMENT T4 – Radio and Electronic Fundamentals

[5 exam questions – 5 groups]

T4A – Names of electrical units, DC and AC, what is a radio signal, conductors and insulators, electrical components

T4B – relationship between frequency and wavelength, identification of bands, names of frequency ranges, types of waves

T4C - How radio works: receivers, transmitters, transceivers, amplifiers, power supplies, types of batteries, service life

T4D – Ohms law relationships

T4E - Power calculations, units, kilo, mega, milli, micro

SUBELEMENT T5 – Station Setup and Operation

[4 exam questions – 4 groups]

T5A - Station hookup – microphone, speaker, headphones, filters, power source, connecting a computer

T5B - Operating controls

T5C – Repeaters; repeater and simplex operating techniques, offsets, selective squelch, open and closed repeaters, linked repeaters

T5D – Recognition and correction of problems, symptoms of overload and overdrive, distortion, over and under modulation, RF feedback, off frequency signals, fading and noise, problems with digital communications links

SUBELEMENT T6 – Communications Modes and Methods

[3 exam questions - 3 groups]

T6A - Modulation modes, descriptions and bandwidth (AM, FM, SSB)

T6B - Voice communications, EchoLink and IRLP

T6C – Non-voice communications - image communications, data, CW, packet, PSK31, Morse code techniques, Q signals

SUBELEMENT T7 – Special Operations

[2 exam questions – 2 groups]

T7A – Operating in the field, radio direction finding, radio control, contests, special event stations

T7B – Satellite operation, Doppler shift, satellite sub bands, LEO, orbit calculation, split frequency operation, operating protocols, AMSAT, ISS communications

SUBELEMENT T8 – Emergency and Public Service Communications

[3 exam questions – 3 groups]

T8A - FCC declarations of an emergency, use of non-amateur equipment and frequencies, use of equipment by unlicensed persons, tactical call signs

T8B - Preparation for emergency operations, RACES/ARES, safety of life and property, using ham radio at civic events, compensation prohibited

T8C - Net operations, responsibilities of the net control station, message handling, interfacing with public safety officials

SUBELEMENT T9 – Radio Waves, Propagation and Antennas

[3 exam questions – 3 groups]

T9A - Antenna types – vertical, horizontal, concept of gain, common portable and mobile antennas, losses with short antennas, relationships between antenna length and frequency, dummy loads

T9B – Propagation, fading, multipath distortion, reflections, radio horizon, terrain blocking, wavelength vs. penetration, antenna orientation

T9C – Feed lines types, losses vs. frequency, SWR concepts, measuring SWR, matching and power transfer, weather protection, feed line failure modes

SUBELEMENT T0 – Electrical and RF Safety

[3 exam questions – 3 groups]

T0A – AC power circuits, hazardous voltages, fuses and circuit breakers, grounding, lightning protection, battery safety, electrical code compliance

T0B – Antenna installation, tower safety, overhead power lines

T0C - RF hazards, radiation exposure, RF heating hazards, proximity to antennas, recognized safe power levels, hand held safety, exposure to others

Subelement T1

FCC Rules

Your Technician exam (element 2) will consist of 35 questions taken from the Technician question pool as prepared by the Volunteer Examiner Coordinator's Question Pool Committee. A certain number of questions are taken from each of the 10 subelements. There will be 5 questions from the FCC Rules subelement shown in this chapter. These questions are divided into 4 groups labeled T1A through T1D.

You'll often see a reference to Part 97 of the Federal Communications Commission Rules set aside in brackets, like this: [97.3(a) (5)]. This tells you where to find the exact wording of the Rules as they relate to that question. You'll find the complete Part 97 Rules on the ARRL Web site at **www.arrl.org/FandES/field/regulations/news/part97/**, and in the *FCC Rule Book* published by the ARRL.

T1A Basis and purpose of the Amateur Radio Service, penalties for unlicensed operation, other penalties, examinations – 1 exam question

T1A01 Who is an amateur operator as defined in Part 97?

A. A person named in an amateur operator/primary license grant in the FCC ULS database
B. A person who has passed a written license examination
C. The person named on the FCC Form 605 Application
D. A person holding a Restricted Operating Permit

A [97.3(a)(1)] The FCC defines an amateur operator as "*A person holding a written authorization to be the control operator of an amateur station.*" This sounds a little circular! The definition is written this way because the rest of Part 97 requires a precise legal definition. Since the only way to get "written authorization" is to pass one or more of the license elements, this means that you must have passed your exam and received your license to be a control operator.

T1A02 What is one of the basic purposes of the Amateur Radio Service as defined in Part 97?

A. To support teaching of amateur radio classes in schools
B. To provide a voluntary noncommercial communications service to the public, particularly in times of emergency
C. To provide free message service to the public
D. To allow the public to communicate with other radio services

B [97.1] In Section 97.1 of its Rules, the FCC describes the basis and purpose of the amateur service. It consists of five principles, which comprise the foundation and rationale for the existence of Amateur Radio in the US. Here's what the Rules say:

§97.1 Basis and Purpose

The rules and regulations in this part are designed to provide an amateur radio service having a fundamental purpose as expressed in the following principles:

(a) Recognition and enhancement of the value of the amateur service to the public as a voluntary noncommercial communication service, particularly with respect to providing emergency communications.

(b) Continuation and extension of the amateur's proven ability to contribute to the advancement of the radio art.

(c) Encouragement and improvement of the amateur service through rules which provide for advancing skills in both the communication and technical phases of the art.

(d) Expansion of the existing reservoir within the amateur radio service of trained operators, technicians, and electronics experts.

(e) Continuation and extension of the amateur's unique ability to enhance international goodwill.

T1A03 What classes of US amateur radio licenses may currently be earned by examination?

A. Novice, Technician, General, Advanced

B. Technician, General, Advanced

C. Technician, General, Extra

D. Technician, Tech Plus, General

C [97.501] Anyone earning a new Amateur Radio license can earn one of three license classes—Technician, General, and Amateur Extra. These vary in degree of knowledge required and frequency privileges granted. The Tech Plus license is not considered a separate class, but an enhancement to the Technician license.

T1A04 Who is a Volunteer Examiner?

A. A certified instructor who volunteers to examine amateur teaching manuals

B. An FCC employee who accredits volunteers to administer amateur license exams

C. An amateur accredited by one or more VECs who volunteers to administer amateur license exams

D. Any person who volunteers to examine amateur station equipment

C A VEC is an organization that has made an agreement with the FCC to coordinate Amateur Radio license examinations by using Volunteer Examiners (VEs). A VE is a licensed Amateur Radio operator who volunteers to help administer amateur license exams.

T1A05 **How long is a CSCE valid for license upgrade purposes?**

A. 365 days
B. Until the current license expires
C. Indefinitely
D. Until two years following the expiration of the current license

A [97.505(a)(6)] Present your valid CSCE at an examination session as documentation that you have passed an examination element. CSCEs are only valid for 365 days.

T1A06 **How many and what class of Volunteer Examiners are required to administer an Element 2 Technician written exam?**

A. Three Examiners holding any class of license
B. Two Examiners holding any class of license
C. Three Examiners holding a Technician Class license
D. Three Examiners holding a General Class license or higher

D [97.509(a)(b)(3)(i)] An exam session requires at least three VEs holding licenses classes higher than that for which the exams are given. To administer a Technician license exam, the VEs must hold a General or Extra class license. VEs holding a Technician class license can participate in administering the exam, but can not be counted as part of the three required VEs.

T1A07 **Who makes and enforces the rules for the Amateur Radio Service in the United States?**

A. The Congress of the United States
B. The Federal Communications Commission
C. The Volunteer Examiner Coordinators
D. The Federal Bureau of Investigation

B [97.5] Part 97 of the Federal Communication Commission's Rules governs the Amateur Radio Service in the United States. It is the FCC that enforces those rules.

T1A08 **What are two of the five fundamental purposes for the Amateur Radio Service?**

A. To protect historical radio data, and help the public understand radio history
B. To aid foreign countries in improving radio communications and encourage visits from foreign hams
C. To modernize radio electronic design theory and improve schematic drawings
D. To increase the number of trained radio operators and electronics experts, and improve international goodwill

D [97.1] See T1A02

T1A09 What is the definition of an amateur radio station?

A. A station in a public radio service used for radio communications
B. A station using radio communications for a commercial purpose
C. A station using equipment for training new broadcast operators and technicians
D. A station in an Amateur Radio Service consisting of the apparatus necessary for carrying on radio communications

D [97.3(a)(5)] The FCC defines an amateur station as, "*A station licensed in the amateur service, including the apparatus necessary for carrying on radio communications.*" Another circular-sounding definition, but remember that the definitions are used in specific legal regulations. What the FCC is saying is that a station that conducts radio communications as required by the amateur service rules in Part 97 meets the definition of an amateur station.

T1A10 What is a transmission called that disturbs other communications?

A. Interrupted CW
B. Harmful interference
C. Transponder signals
D. Unidentified transmissions

B [97.3(A)(23)] A transmission that disturbs other authorized communications is called harmful interference. FCC Rules define harmful interference as, "*Interference which endangers the functioning of a radionavigation service or of other safety communication service operating in accordance with the Radio Regulations.*" Not all interference is harmful.

T1B ITU regions, international regulations, US call sign structure, special event calls, vanity call signs - 1 exam question

T1B01 **What is the ITU?**

A. The International Telecommunications Utility
B. The International Telephone Union
C. The International Telecommunication Union
D. The International Technology Union

C [97.3(a)(28)] The International Telecommunications Union (ITU) is an international body of the United Nations that has responsibility for organizing the various radio services on a worldwide basis. This includes arranging international telecommunications treaties, as well as administrative responsibilities such as call signs and frequency allocations.

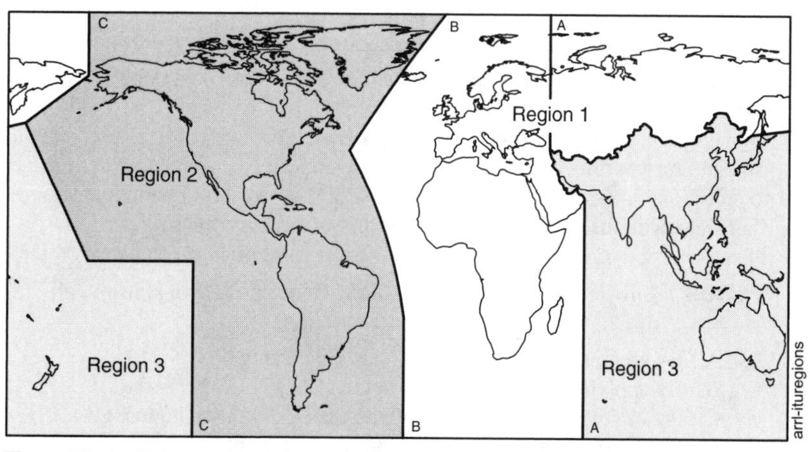

Figure T1-1 - This map shows the world divided into three International Telecommunications Union (ITU) Regions. The US is part of Region 2.

T1B02 **What is the purpose of ITU Regions?**

A. They are used to assist in the management of frequency allocations
B. They are useful when operating maritime mobile
C. They are used in call sign assignments
D. They must be used after your call sign to indicate your location

A [97.301] In order to make administering the world's telecommunications services easier, the ITU divided the world into three regions as shown in Fig T1-1. ITU Region 1 comprises Africa Europe, Russia, and parts of the Middle East. ITU Region 2 comprises North and South American as well as the Caribbean Islands, Hawaii, and Alaska. By allowing each region to use different sets of regulations, it is easier to develop telecommunications services.

T1B03 What system does the FCC use to select new amateur radio call signs?

A. Call signs are assigned in random order
B. The applicant is allowed to pick a call sign
C. Call signs are assigned in sequential order
D. Volunteer Examiners choose an unassigned call sign

C [97.17(d)] The FCC issues call signs on a sequential basis. When your application is processed, you get the next call sign in order of ITU prefix letter(s), call sign district numeral, and suffix in strict alphabetic order.

T1B04 What FCC call sign program might you use to obtain a call sign containing your initials?

A. The vanity call sign program
B. The sequential call sign program
C. The special event call sign program
D. There is no FCC provision for choosing a your call sign

A [97.19(d)] In addition to the automatic sequential call sign system, the FCC has a way for Amateurs to select their own call sign. This is called the Vanity Call Sign program. Vanity call sign choices must have the same format as a sequential call sign for your license class. As a Technician licensee, you would be eligible to select any unassigned "one-by-three" or a "two-by-three" format call sign. Call signs with the operator's initials for the suffix are popular.

T1B05 How might an amateur radio club obtain a club station call sign?

A. By applying directly to the FCC in Gettysburg, PA
B. By applying through a Club Station Call Sign Administrator
C. By submitting a FCC Form 605 to the FCC in Washington, DC
D. By notifying a VE team using NCVEC Form 605

B [97.17(b)(2)] Amateur Radio clubs can also obtain a station call sign. The club must name one member as the license trustee, to have primary responsibility for the license. The club must apply for the license through an FCC-approved Club Station Call Sign Administrator. The Call Sign Administrator collects the required information and sends it to the FCC. The ARRL/VEC, W4VEC, and W5YI-VEC are Club Station Call Sign Administrators.

T1B06 Who is eligible to apply for temporary use of a 1-by-1 format Special Event call sign?

A. Only Amateur Extra class amateurs
B. Only military stations
C. Any FCC-licensed amateur
D. Only trustees of amateur radio club stations

C Individuals or club groups who plan to operate an amateur station to commemorate a special event can apply for a Special-Event Call Sign. These special "one-by-one" format call signs can help call attention to the on-the-air operation at the special event. These call signs are issued for a short-term operation, normally 15 days or less. Any licensed amateur is allowed to apply for a Special-Event Call Sigh. The FCC-approved Special-Event Call Sign Administrators coordinate these call signs. The ARRL/VEC, Laurel (Maryland) Amateur Radio Club, W4VED, and W5YI-VEC are Special-Event Call Sign Administrators.

T1B07 When are you allowed to operate your amateur station in a foreign country?

A. When there is a reciprocal operating agreement between the countries
B. When there is a mutual agreement allowing third party communications
C. When authorization permits amateur communications in a foreign language
D. When you are communicating with non-licensed individuals in another country

A [97.107] A reciprocal operating agreement between the US and the foreign country enables you to operate your radio according to the rules of that country. There are three types of agreement; the International Amateur Radio Permit, the European Conference of Postal and Telecommunications Administration (CEPT) agreement, and International Reciprocating Agreements between the foreign country and the US.

T1B08 Which of the following call signs is a valid US amateur call?

A. UZ4FWD
B. KBL7766
C. KB3TMJ
D. VE3TWJ

C The first letter of a US call will always be A, K, N, or W. These letters are assigned to the United States for all call-sign prefixes. US amateur call signs begin with one or two letters, followed by a single numeral, then one to three letters. The initial letters and the numeral make up the prefix and the following letters make up the suffix. Other countries use different prefixes—LA2UA is a Norwegian call sign, VE3BKJ is from Canada, and VU2HO is from India.

T1B09 What letters must be used for the first letter in US amateur call signs?

A. K, N, U and W
B. A, K, N and W
C. A, B, C and D
D. A, N, V and W

B See T1B08

T1B10 What numbers are used in US amateur call signs?

A. Any two-digit number, 10 through 99
B. Any two-digit number, 22 through 45
C. A single digit, 1 though 9
D. A single digit, 0 through 9

D A single numeral is used in US call signs to indicate the district in which the call was first issued. Every US amateur call sign includes a single-digit number, 0 through 9, corresponding to the call districts shown in the figure. Amateurs may keep their call sign when moving from one district to another so the number in the call is not always an indicator of where an amateur is located.

Figure T1-2 - The 10 US call districts. An amateur holding the call sign K1STO lived in the first district when assigned that call by the FCC. Alaska is part of the seventh call district, but has its own set of prefixes: AL7, KL7, NL7 and WL7. Hawaii, part of the sixth district, has the AH6, KH6, NH6 and WH6 prefixes.

T1C Authorized frequencies (Technician), reciprocal licensing, operation near band edges, spectrum sharing – 1 exam question

T1C01 What is required before you can control an amateur station in the US?

A. You must hold an FCC restricted operator's permit for a licensed radio station
B. You must submit an FCC Form 605 with a license examination fee
C. You must be named in the FCC amateur license database, or be an alien with reciprocal operating authorization
D. The FCC must issue you a Certificate of Successful Completion of Amateur Training

C [97.5(a)] - As long as you are listed in the FCC's database of licensed amateurs, you are authorized to be a control operator of an amateur station. To get into the database, it's necessary to have passed the exam or exams required for a license. Foreign citizens with a ham license whose countries have established a reciprocal licensing authorization (see Section 97.107) can also be control operators.

T1C02 Where does a US amateur license allow you to transmit?

A. From anywhere in the world
B. From wherever the Amateur Radio Service is regulated by the FCC or where reciprocal agreements are in place
C. From a country that shares a third party agreement with the US
D. Only from the mailing address printed on your license

B [97.5(a)] - A US amateur license allows you to operate wherever the FCC regulates the amateur service. That includes the 50 states as well as any territories under US government control. You may also operate from a foreign country if that country has a reciprocal operating agreement with the US as defined by Section 97.107.

T1C03 Under what conditions are amateur stations allowed to communicate with stations operating in other radio services?

A. When other radio services make contact with amateur stations
B. When authorized by the FCC
C. When communicating with stations in the Family Radio Service
D. When commercial broadcast stations are off the air

B [97.111] - Under most circumstances you are not permitted to contact stations in other radio services—that's the whole point of having different services! However, when authorized by the FCC or during a communications emergency, you may contact those stations. Section 97.403 defines a communications emergency as "*communication needs in connection with the immediate safety of human life and immediate protection of property when normal communications systems are not available.*"

T1C04 Which frequency is within the 6-meter band?

A. 49.00 MHz
B. 52.525 MHz
C. 28.50 MHz
D. 222.15 MHz

B [97.301(a)] - The 6-meter band extends from 50.0 to 54.0 MHz in ITU Region 2, which includes North and South America.

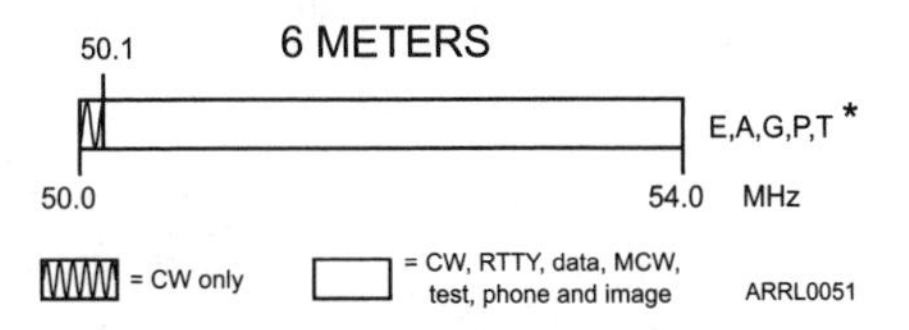

T1C05 Which amateur band are you using when transmitting on 146.52 MHz?

A. 2 meter band
B. 20 meter band
C. 14 meter band
D. 6 meter band

A [97.301(a)] - The 2-meter band extends from 144.0 to 148.0 MHz in ITU Region 2, which includes North and South America.

144.1 2 METERS

E,A,G,P,T

144.0 148.0 MHz

= CW only = CW, RTTY, data, MCW, test, phone and image

ARRL0052

E = Extra Class
A = Advanced
G = General
P = Technician Plus or Technician who has passed 5 wpm Morse Code Exam
T = Technician

T1C06 Which 70-centimeter frequency is authorized to a Technician class license holder operating in ITU Region 2?

A. 455.350 MHz
B. 146.520 MHz
C. 443.350 MHz
D. 222.520 MHz

C [97.301(a)] - The 70-centimeter band extends from 420.0 to 450.0 MHz in ITU Region 2, which includes North and South America.

70 CENTIMETERS **

E,A,G,P,T

420.0 450.0 MHz

**Geographical and power restrictions apply to all bands with frequencies above 420 MHz. See *The ARRL's FCC Rule Book* for more information about your area.

= CW, RTTY, data, MCW, test, phone and image

ARRL0054

T1C07 Which 23-centimeter frequency is authorized to a Technician class license holder operating in ITU Region 2?

A. 2315 MHz
B. 1296 MHz
C. 3390 MHz
D. 146.52 MHz

B [97.301(a)] - The 23-centimeter band extends from 1240 to 1300 MHz in ITU Region 2, which includes North and South America.

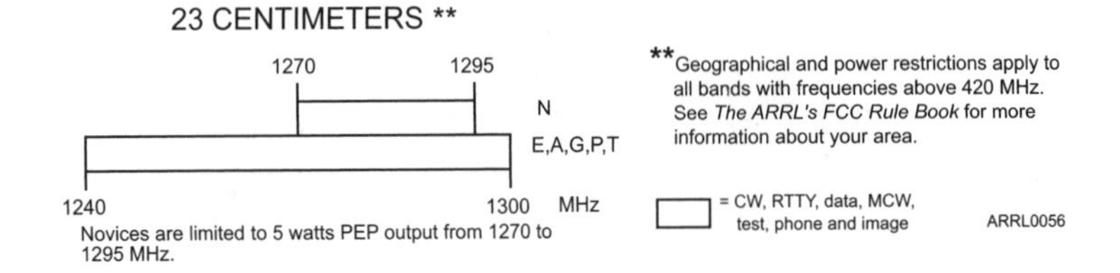

T1C08 What amateur band are you using if you are operating on 223.50 MHz?

A. 15 meter band
B. 10 meter band
C. 2 meter band
D. 1.25 meter band

D [97.301(a)] - The 1.25-meter band extends from 222 to 225 MHz in ITU Region 2, which includes North and South America.

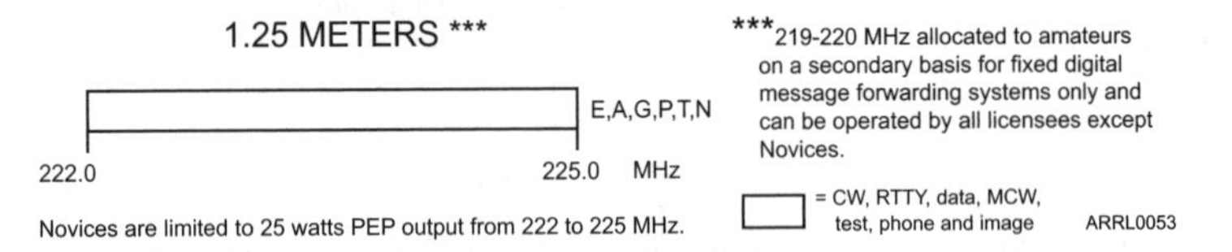

T1C09 What do the FCC rules mean when an amateur frequency band is said to be available on a secondary basis?

A. Secondary users of a frequency have equal rights to operate
B. Amateurs are only allowed to use the frequency at night
C. Amateurs may not cause harmful interference to primary users
D. Secondary users are not allowed on amateur bands

C [97.303] - A radio service that is designated as the primary service on a band is protected from interference caused by other radio services. A radio service that is designated as the secondary service must not cause harmful interference to, and must accept interference from, stations in a primary service. The amateur service has many different frequency bands. Some of them are allocated on a primary basis and some are secondary.

T1C10 When may a US amateur operator communicate with an amateur in a foreign country?

A. Only when a third-party agreement exists between the US and the foreign country
B. At any time except between 146.52 and 146.58 MHz
C. Only when a foreign amateur uses English
D. At any time unless prohibited by either government

D [97.111] - You may converse with any amateur station at any time. This includes amateurs in foreign countries, unless either amateur's government prohibits the communications. (There are times when a government will not allow it amateurs to talk with people in other countries.)

T1C11 Which of the following types of communications are not permitted in the Amateur Radio Service?

A. Brief transmissions to make adjustments to the station
B. Brief transmissions to establish two-way communications with other stations
C. Transmissions to assist persons learning or improving proficiency in CW
D. Communications on a regular basis that could reasonably be furnished alternatively through other radio services

D [97.113(a)(5)] - The FCC has created the different radio services and allocated spectrum to them to satisfy specific communication needs. While the amateur service occasionally provides communications that are similar to those of other services, it is not intended for those communications to take place on a regular basis.

T1D The station license, correct name and address on file, license term, renewals, grace period – 1 exam question

T1D01 Which of the following services are issued an operator station license by the FCC?

A. Family Radio Service
B. Amateur Radio Service
C. General Radiotelephone Service
D. The Citizens Radio Service

B [97.17(a)] - The FCC has created several unlicensed services so that citizens have access to wireless technology without the requirement of passing an examination or paying a fee. However, the purpose and use of these services is tightly constrained, including restrictions on experimentation and station construction.

T1D02 Who can become an amateur licensee in the US?

A. Anyone except a representative of a foreign government
B. Only a citizen of the United States
C. Anyone except an employee of the US government
D. Anyone

A 97.5(b)(1)] - Amateur Radio is open to almost everyone. Anyone, except an agent or representative of a foreign government, is eligible to qualify for an Amateur Radio operator license.

T1D03 What is the minimum age required to hold an amateur license?

A. 14 years or older
B. 18 years or older
C. 70 years or younger
D. There is no minimum age requirement

D [97.5(b)(1)] - there is no age requirement for individuals to hold an Amateur Radio operator license. You can't be too young or too old to qualify for a license.

T1D04 What government agency grants your amateur radio license?

A. The Department of Defense
B. The Bureau of Public Communications
C. The Department of Commerce
D. The Federal Communications Commission

D [97.5(a)] - In the United States, the Federal Communications Commission issues amateur licenses. A US amateur license allows you to operate wherever the FCC regulates the amateur service.

T1D05 How soon may you transmit after passing the required examination elements for your first amateur radio license?

A. Immediately
B. 30 days after the test date
C. As soon as your license grant appears in the FCC's ULS database
D. As soon as you receive your license in the mail from the FCC

C [97.5(a)] - Your Amateur Radio license is valid as soon as the FCC grants the license and posts the information about your license in their database. You don't have to wait for the actual printed license to arrive by mail before you begin to transmit! You can check the FCC database on one of the Internet servers or someone else can check it for you. The ARRL maintains a call sign lookup service at **www.arrl.org/fcc/fcclook.php**.

T1D06 What is the normal term for an amateur station license grant?

A. 5 years
B. 7 years
C. 10 years
D. For the lifetime of the licensee

C [97.25(a)] - The FCC issues all amateur licenses for a 10-year term. You should always renew your license for another 10-year term before it expires.

T1D07 What is the grace period during which the FCC will renew an expired 10-year license without re-examination?

A. 2 years
B. 5 years
C. 10 years
D. There is no grace period

A [97.21(b)] - If you do forget to renew your license, you have up to two years to apply for renewal without having to take the exams again. After the two-year grace period, you will have to retake the exams. Your license is not valid during this two-year grace period, however, and you may not operate an amateur station with an expired license. All that the grace period implies is that the FCC will renew the license if you apply during that period.

T1D08 What is your responsibility as a station licensee?

A. You must allow another amateur to operate your station upon request
B. You must be present whenever the station is operated
C. You must notify the FCC if another amateur acts as the control operator
D. Your station must be operated in accordance with the FCC rules

D [97.103(a)] - Part 97 of the FCC Rules provides quite a bit of guidance about what you can and cannot do as an Amateur Radio operator. Of course every possible situation can't be covered, so you will have to use good judgement in your operating practices. You should be familiar with the specific requirements of the Rules and always try to operate within the intent of those Rules. As an Amateur Radio operator, you are always responsible for the proper operation of your Amateur Radio station.

T1D09 **When may the FCC revoke or suspend a license if the mailing address of the holder is not current with the FCC?**

A. If mail is returned to the FCC as undeliverable
B. When the licensee transmits without having updated the address
C. When the licensee operates portable at a different address
D. If the address is not updated within the 2 year grace period

A [97.23] - You are required to maintain on file with the FCC a mailing address in an area where the FCC regulates the amateur service and where you can receive mail delivery by the United States Postal Service. If the FCC needs to contact you, this is how they will do so—by mail. If you do not maintain a current address and FCC mail to you is returned to them as undeliverable, they will have no way of contacting you and may revoke your license.

T1D10 **The FCC requires which address to be kept up to date on the Universal Licensing System database?**

A. The station location address
B. The station licensee mailing address
C. The station location address and mailing address
D. The station transmitting location address

B [97.23] - See T1D09

T1D11 **When are you permitted to continue to transmit if you forget to renew your amateur license and it expires?**

A. Transmitting is not allowed until the license is renewed and appears on the FCC ULS database
B. When you identify using the suffix EXP
C. When you notify the FCC you intend to renew within 90 days
D. Transmitting is allowed any time during the 2-year grace period

A [97.21(b)] - You may not transmit while your amateur license is expired. You must renew your license before you may transmit. (See also T1D07)

T1D12 **Why must an Amateur radio operator have a correct name and mailing address on file with the FCC?**

A. To receive mail delivery from the FCC by the United States Postal Service
B. So the FCC Field office can contact the licensee
C. It isn't required when you haven't operated your station in a year
D. So the FCC can locate your transmitting location

A [97.23] - Mail is the method by which the FCC will communicate with you. (See also question T1D09)

Subelement **T2**

FCC Rules Continued

Your Technician exam (element 2) will consist of 35 questions taken from the Technician question pool as prepared by the Volunteer Examiner Coordinator's Question Pool Committee. A certain number of questions are taken from each of the 10 subelements. There will be 4 questions from the FCC Rules subelement shown in this chapter. These questions are divided into 4 groups labeled T2A through T2D.

You'll often see a reference to Part 97 of the Federal Communications Commission Rules set aside in brackets, like this: [97.3(a) (5)]. This tells you where to find the exact working of the Rules as they relate to that question. You'll find the complete Part 97 Rules on the ARRL Web site at **www.arrl.org/FandES/field/regulations/news/part97/**, and in the *FCC Rule Book* published by the ARRL.

T2A Prohibited communications: music, broadcasting, codes and ciphers, business use, permissible communications, bulletins, code practice, incidental music – 1 exam question

T2A01 When is an amateur station authorized to transmit information to the general public?

A. Never
B. Only when the operator is being paid
C. Only when the transmission lasts more than 10 minutes
D. Only when the transmission lasts longer than 15 minutes

A [97.113(b)] - Amateur Radio is a two-way communications service. Amateur Radio stations may not engage in broadcasting. (see question T2A06)

T2A02 When is an amateur station authorized to transmit music?

A. Amateurs may not transmit music, except as incidental to an authorized rebroadcast of space shuttle communications
B. Only when the music produces no spurious emissions
C. Only to interfere with an illegal transmission
D. Only when the music is above 1280 MHz

A [97.113(a)(4), 97.113(e)] - Under FCC Rules, amateurs may not transmit music of any form. This means you can't transmit your band's practice session or play the piano for transmission over the air. Retransmitting music from a broadcast program on a radio in your car or shack is also prohibited--so turn the radio down when you're on the air! There is one exception to the "No Music" rule. If you obtain special permission from NASA to retransmit the audio from a space shuttle mission or the International Space Station for other amateurs to listen, and during that retransmission NASA or the astronauts play some music over the air, that's OK.

T2A03 When is the transmission of codes or ciphers allowed to hide the meaning of a message transmitted by an amateur station?

A. Only during contests
B. Only when operating mobile
C. Only when transmitting control commands to space stations or radio control craft
D. Only when frequencies above 1280 MHz are used

C [97.113(a)(4), 97.211(b), 97.217] - You can't use codes or ciphers (also known as *encryption*) to obscure the meaning of transmissions. This means you can't make up a "secret" code to send messages over the air to a friend. However, there are special exceptions. Control signals transmitted for remote control of model craft are not considered codes or ciphers. Neither are telemetry signals, such as a satellite might transmit to report on internal conditions. A space station (satellite) control operator may use specially coded signals to control the satellite.

T2A04 When may an amateur station transmit false or deceptive signals?

A. Never
B. When operating a beacon transmitter in a "fox hunt" exercise
C. Only when making unidentified transmissions
D. When needed to hide the meaning of a message for secrecy

A [97.113(a)(4)] - Amateurs (and operators using other radio services, as well) may not transmit false or deceptive signals, such as a distress call when no emergency exists. You must not, for example, start calling MAYDAY (an international distress signal) unless you are in a life-threatening situation. The FCC and law enforcement agencies will treat such behavior as a serious violation of the law.

T2A05 **When may an amateur station transmit unidentified communications?**

A. Only during brief tests not meant as messages
B. Only when they do not interfere with others
C. Only when sent from a space station or to control a model craft
D. Only during two-way or third party communications

C [97.119(b)] - The rules prohibit unidentified communications or signals in the amateur service--signals where the transmitting station's call sign is not included. There are two exceptions to this rule which exempt space stations and telecommand stations from having to identify their transmissions.

T2A06 **What does the term broadcasting mean?**

A. Transmissions intended for reception by the general public, either direct or relayed
B. Retransmission by automatic means of programs or signals from non-amateur stations
C. One-way radio communications, regardless of purpose or content
D. One-way or two-way radio communications between two or more stations

A [97.3(a)(10)] - For the purposes of radio regulation, broadcasting means the transmission of information intended for reception by the general public. These broadcast transmissions may either be direct or relayed. Rules for Amateur Radio, a two-way communications service, prohibit broadcasting by amateur stations.

T2A07 **Which of the following are specifically prohibited in the Amateur Radio Service?**

A. Discussion of politics
B. Discussion of programs on broadcast stations
C. Indecent and obscene language
D. Morse code practice

C [97.113(a)(4)] - Amateurs may not use obscene or indecent language—it is prohibited by the Rules. Amateur radio transmissions are public and anyone of any age can hear them. Depending on conditions and frequency, they might be heard anywhere in the world. There is no list of "words you can't say on Amateur Radio", but bear in mind the public nature of our communications and avoid any questionable language.

T2A08 Which of the following one-way communications may not be transmitted in the Amateur Radio Service?

A. Telecommand of model craft
B. Broadcasts intended for reception by the general public
C. Brief transmissions to make adjustments to the station
D. Morse code practice

B [97.3(a)(10), 97.113(b)] - Amateur Radio is a two-way communications service. Nevertheless, there are exceptions that allow amateurs to make one-way transmissions to control model craft, make adjustments to their stations, and even to transmit bulletins and Morse code practice. Amateurs are not permitted to engage in broadcasting as discussed for questions T2A01 and T2A06.

T2A09 When does the FCC allow an amateur radio station to be used as a method of communication for hire or material compensation?

A. Only when making test transmissions
B. Only when news is being broadcast in times of emergency
C. Only when in accordance with Part 97 rules
D. Only when your employer is using amateur radio to broadcast advertising

C [97.113(2)] - In nearly all instances, amateurs are prohibited from receiving compensation for operating an amateur station, such as when you are providing communications at public events like race or parade. The prohibition also includes operating as part of your regular job. For example, if you work at a hospital, it would be acceptable for you to be assigned the job of building an emergency amateur station. However, you could not be assigned to operate the station as part of your regular duties. An exception is made for teachers, who are permitted to operate an amateur station as part of their classroom instruction. Another exception is made for a club that employs an amateur to operate the club station to send bulletins and other information of interest to amateurs. (See question T2D11.)

T2A10 What types of communications are prohibited when using a repeater autopatch?

A. Calls to a recorded weather report
B. Calls to your employer requesting directions to a customer's office
C. Calls to the police reporting a traffic accident
D. Calls to a public utility reporting an outage of your telephone

B [97.113(a)(3),(a)5(e)] - FCC rules prohibit amateur operators from engaging in any type of business communications. You may not conduct communications for your own business, or for your employer. This includes using the autopatch on your local repeater.

T2A11 **When may you use your station to tell people about equipment you have for sale?**

A. Never
B. When you are conducting an on-line auction
C. When you are offering amateur radio equipment for sale or trade on an occasional basis
D. When you are helping a recognized charity

C [97.113(a)3] - You are permitted to let other amateurs know that you have amateur radio equipment for sale or trade. This is a common practice on local "swap-and-shop" nets. The equipment or materials must be normally used in amateur radio--no skis, cars, record collections, or antiques! You are also prohibited from selling or trading on a regular basis, particularly if you make a profit on the transaction.

T2B Basic identification requirements, repeater ID standards, identification for non-voice modes, identification requirements for mobile and portable operation – 1 exam question

T2B01 **What must you transmit to identify your amateur station?**

A. Your tactical ID
B. Your call sign
C. Your first name and your location
D. Your full name

B [97.119(a)] - The FCC Rules require that you transmit your call sign to identify your station. You only have to identify (or ID) at the end of a QSO and at least once every 10 minutes during its course. There is no legal requirement to transmit your call sign at the beginning of the contact or the call sign of stations you're in contact with at any time.

T2B02 **What is a transmission called that does not contain a station identification?**

A. Unidentified communications or signals
B. Reluctance modulation
C. Test emission
D. Intentional interference

A [97.119(a)] - Unidentified communications or signals are signals where the transmitting station's call sign is not included. Be sure you understand the proper station identification procedures, so you don't violate this rule.

T2B03 **How often must an amateur station transmit the assigned call sign?**

A. At the beginning of each transmission and every 10 minutes during communication
B. Every 10 minutes during communications and at the end of each communication
C. At the end of each transmission
D. Only at the end of the communication

B [97.119(a)] - FCC regulations are very specific about station identification. You must identify your station every ten minutes (or more frequently) during a contact and at the end of the contact. (See also question T2B01.)

T2B04 **What is an acceptable method of transmitting a repeater station identification?**

A. By phone using the English language
B. By video image conforming to applicable standards
C. By Morse code at a speed not to exceed 20 words per minute
D. All of these answers are correct.

D [97.119(b)] - A repeater, operating under remote or automatic control, must identify its transmissions. The FCC provides three methods of identifying; spoken call signs, Morse code, or in the case of a video repeater used for amateur television (ATV) by a video image.

T2B05 **What identification is required when two amateur stations end communications?**

A. No identification is required
B. One of the stations must transmit both stations' call signs
C. Each station must transmit its own call sign
D. Both stations must transmit both call signs

C [97.119(a)] - You are required to transmit your call sign at the end of a contact and every 10 minutes during the contact. There is no requirement to give the call sign of the station or stations with whom you're in contact either at the beginning or end of the contact. It is usually good practice to give the call signs of the other stations, particularly when establishing contact, but it is not required.

T2B06 **What is the longest period of time an amateur station can operate without transmitting its call sign?**

A. 5 minutes
B. 10 minutes
C. 15 minutes
D. 30 minutes

B [97.119(a)] - Regardless of where you are operating and whether from a mobile, portable, or home station, you must identify your station at least once every 10 minutes. (See also questions T2A01 and T2A06.)

T2B07 What is a permissible way to identify your station when you are speaking to another amateur operator using a language other than English?

A. You must identify using the official version of the foreign language
B. Identification is not required when using other languages
C. You must identify using the English language
D. You must identify using phonetics

C [97.119(b)(2)] - You can use any language you want to communicate with other amateurs, even Esperanto! Amateur Radio provides you a great opportunity to practice your foreign language skills with other hams that speak that language. When you give your identification as a US station, however, you must use English. Foreign stations may have a similar requirement to use their native language when identifying their transmissions.

T2B08 How often must you identify using your assigned call sign when operating while using a special event call sign?

A. Every 10 minutes
B. Once when the event begins and once when it concludes
C. Never
D. Once per hour

D [97.119(d)] - During the period when you are authorized to use a Special Event call sign, such as W1A or K5Z, you must identify using the special call sign just as if you are when using your own or your club's call sign. However, once per hour, you must transmit the call sign of the individual or club that is authorized to use the special call.

T2B09 What is required when using one or more self-assigned indicators with your assigned call sign?

A. The indicator must not conflict with an indicator specified by FCC rules or with a prefix assigned to another country
B. The indicator must consist only of numeric digits
C. The indicator must include the 2-letter abbreviation for your state
D. The indicator must be separated from your call sign by a double slash mark

A [97.119(c)] - An indicator is words, letters, or numerals appended to and separated from your call sign during identification. For instance, you might add the word "mobile" to your call sign when you operating from your car on your way to work. Indicators are used when you are operating away from your home station, such as a portable operation or from a different country. If you operate from Hawaii on a vacation, you would add KH6 before or after your call sign and separated from it by saying "portable" or sending a slash on CW. The indicator can not conflict with an indicator specified by FCC Rules or with a prefix assigned to another country.

T2B10 **What is the correct way to identify when visiting a station if you hold a higher-class license than that of the station licensee and you are using a frequency not authorized to his class of license?**

A. Send your call sign first, followed by his call sign
B. Send his call sign first, followed by your call sign
C. Send your call sign only, his is not required
D. Send his call sign followed by "/KT"

B [97.119(e)] - Since you are using your host's station, you must begin your identification with the host's call sign. To show that the control operator of the station (that's you) is authorized to transmit on the frequency, your call sign is added, following a slash on CW or words such as "operated by" on phone. The intent is to let any listener determine whether the station has authorization to operate on the frequency.

T2B11 **When exercising the operating privileges earned by examination upgrade of a license what is meant by use of the indicator "/AG"?**

A. Authorized General
B. Adjunct General
C. Address as General
D. Automatically General

A [97.119(f)(2)] - Adding these indicators shows to the listener that you have passed an exam and have requested that your license class be upgraded. In this case, you have passed the necessary elements to upgrade from Novice or Technician to General Class. By passing the exam, you are authorized to use the new privileges, but need to add the indicator to inform listening stations.

T2C Definition of control operator, location of control operator, automatic and remote control, auxiliary stations – 1 exam question

T2C01 **What must every amateur station have when transmitting?**

A. A frequency-measuring device
B. A control operator
C. A beacon transmitter
D. A third party operator

B [97.7] - All amateur transmissions must be made under the supervision of a control operator whose job it is to be sure FCC Rules are followed. The control operator may not have to be physically present at the transmitter (called *remote control*) and may also set up a station to be controlled by devices or procedures that assure the rules are followed (called *automatic control*). Nevertheless, a control operator must take responsibility for the station's proper operation. (See also question T2C07.)

T2C02 How many amateur operator / primary station licenses may be held by one person?

A. As many as desired
B. One for each portable transmitter
C. Only one
D. One for each station location

C [97.5(b)(1)] - Because the amateur license is both an operator license and your primary station license, you can only hold one at a time. You are allowed to construct as many stations as you want, but there is only one of you to operate them!

T2C03 What minimum class of amateur license must you hold to be a control operator of a repeater station?

A. Technician Plus
B. Technician
C. General
D. Amateur Extra

B [97.205(a)] - A holder of a Technician, General, Advanced, or Amateur Extra Class license may also be the control operator of a repeater, if the repeater is operating using privileges available to that license. For example, a Technician-class licensee can be the control operator of a 2-meter repeater, but not a 10-meter repeater.

T2C04 Who is responsible for the transmissions from an amateur station?

A. Auxiliary operator
B. Operations coordinator
C. Third-party operator
D. Control operator

D [97.3(a)(1)(2)] - The FCC Rules specifically state that the control operator for an amateur station is the person designated to be responsible for the transmissions from that station to assure compliance with the FCC Rules. This is also true for repeater stations, although the operator using the repeater is also responsible for proper operation.

T2C05 When must an amateur station have a control operator?

A. Only when training another amateur
B. Whenever the station receiver is operated
C. Whenever the station is transmitting
D. A control operator is not needed

C [97.7] - All amateur transmissions must be made under the supervision of a control operator whose job it is to be sure FCC Rules are followed. This is true even if the station is operating under automatic control. (See also questions T2C01 and T2C07.)

T2C06 What is the control point of an amateur station?

A. The on/off switch of the transmitter
B. The input/output port of a packet controller
C. The variable frequency oscillator of a transmitter
D. The location at which the control operator function is performed

D [97.3] - This sounds like a circular definition, but the rule allows for an amateur station to be controlled from a remote location or even under automated control by devices and procedures that assure proper operation. (See also question T2C07.)

T2C07 What type of amateur station does not require a control operator to be at the control point?

A. A locally controlled station
B. A remotely controlled station
C. An automatically controlled station
D. An earth station controlling a space station

C [97.109(d)] - A station operating under automatic control does not require the control operator to be present at the control point. Automatic control is defined by the FCC as "*The use of devices and procedures for control of a station when it is transmitting so that compliance with the FCC Rules is achieved without the control operator being present at a control point.*" This is "hands off" operation; there's nobody at the control point, *but there must still be a control operator available who is responsible*. Only stations specifically designated in Part 97 may be automatically controlled.

T2C08 What are the three types of station control permitted and recognized by FCC rule?

A. Local, remote and automatic control
B. Local, distant and automatic control
C. Remote, distant and unauthorized control
D. All of the choices are correct

A [97.3(a)] - Local control is performed when the control point and the control operator are both physically located at the transmitter. This is the most common form of control. Under remote control the control operator indirectly manipulates the transmitter through a control link that could operate via telephone or radio. Automatic control involves the use of devices and procedures that control the transmitter.

T2C09 What type of control is being used on a repeater when the control operator is not present?

A. Local control
B. Remote control
C. Automatic control
D. Uncontrolled

C [97.3(a)] - See question T2C07

T2C10 **What type of control is being used when transmitting using a handheld radio?**

A. Radio control
B. Unattended control
C. Automatic control
D. Local control

D [97.109(a)] - Because the operator is directly manipulating the transmitter, this is local control.

T2C11 **What type of control is used when the control operator is not at the station location but can still make changes to a transmitter?**

A. Local control
B. Remote control
C. Automatic control
D. Uncontrolled

B [97.3] - Because the transmitter is manipulated by the control operator through a control link, this is remote control.

T2C12 **What is the definition of a control operator of an amateur station?**

A. Anyone who operates the controls of the station
B. Anyone who is responsible for the station's equipment
C. An operator designated by the licensee to be responsible for the station's transmissions to assure compliance with FCC rules
D. The operator with the highest class of license who is in control of the station

C [97.3(a)(13)] - The control operator for an amateur station is the person designated to be responsible for the transmissions from that station to assure compliance with the FCC Rules.

T2D Operating another person's station, guest operators at your station, third party communications, autopatch, incidental business use, compensation of operators, club stations, station security, station inspection, protection against unauthorized transmissions – 1 exam question

T2D01 Who is responsible for proper operation if you transmit from another amateur's station?

A. Both of you
B. Only the other station licensee
C. Only you as the control operator
D. Only the station licensee, unless the station records shows another control operator at the time

A [97.103(a)] - Any amateur may designate another licensed operator as the control operator. However, the FCC holds both the control operator and the station licensee responsible for proper operation of the station.

T2D02 What operating privileges are allowed when another amateur holding a higher class license is controlling your station?

A. All privileges allowed by the higher class license
B. Only the privileges allowed by your license
C. All the emission privileges of the higher class license, but only the frequency privileges of your license
D. All the frequency privileges of the higher class license, but only the emission privileges of your license

A [97.105(b)] - If you let another amateur with a higher-class license than yours control your station, any privileges allowed by his or her license are permitted. They're permitted as long as proper identification procedures are followed (see question T2B10.)

T2D03 What operating privileges are allowed when you are the control operator at the station of another amateur who has a higher class license than yours?

A. Any privileges allowed by the higher class license
B. Only the privileges allowed by your license
C. All the emission privileges of the higher class license, but only the frequency privileges of your license
D. All the frequency privileges of the higher class license, but only the emission privileges of your license

B [97.105(a)] - Suppose you hold a Technician class license and are the control operator at the station of another amateur with a higher-class license than yours (a General, Advanced or Amateur Extra class.) In this situation, when you are responsible for the station's operation, you can use only the privileges allowed by your license!

T2D04 Which of the following is a prohibited amateur radio transmission?

A. Using amateur radio to seek emergency assistance
B. Using amateur radio for conducting business
C. Using an amateur phone patch to call for a taxi or food delivery
D. Using an amateur phone patch to call home to say you are running late

B [97.113(a)(3)] - FCC Rules prohibit amateur operators from engaging in any type of business communications. You may not conduct communications for your own business, or for your employer. This includes using a phone patch at a home station or a repeater's autopatch.

T2D05 What is the definition of third-party communications?

A. A message sent between two amateur stations for someone else
B. Public service communications for a political party
C. Any messages sent by amateur stations
D. A three-minute transmission to another amateur

A [97.3(a)(46)] - A message sent between two amateur station for someone else who is not a licensed amateur is third-party communications. (Hams call it *third-party traffic*.) For example, sending a message from your mother-in-law to her relatives on Valentine's Day is third-party communications. The message need not be written and the third-party can be present at your station, such as allowing a visiting student from a foreign country to speak to his parents via your station.

T2D06 How many persons are required to be members of a club for a club station license to be issued by the FCC?

A. At least 5
B. At least 4
C. A trustee and 2 officers
D. At least 2

B [97.5(b)(2)] - In order to insure that a club actually exists, the FCC requires that it have at least four members, a name, be formally organized and managed, and have a primary purpose devoted to Amateur Radio.

T2D07 When may you operate your amateur station aboard an aircraft?

A. At any time
B. Only while the aircraft is on the ground
C. Only with the approval of the pilot in command and not using the aircraft's radio equipment
D. Only when you have written permission from the airline and only using the aircraft's radio equipment

C [97.11(a)] - Because ships and aircraft operate under rules controlled by other federal agencies, amateur stations may only be operated with the approval of the responsible individual for that ship or aircraft.

T2D08 **When is the FCC allowed to inspect your station equipment and station records?**

A. Only on weekends
B. At any time upon request
C. Never
D. Only during daylight hours

B [97.103(c)] - As a station licensee, your responsibilities include the requirement to make the station and the station records available for inspection upon request by an FCC representative. An inspection can occur at any time, although they are actually quite rare.

T2D09 **How might you best keep unauthorized persons from using your amateur station?**

A. Disconnect the power and microphone cables when not using your equipment
B. Connect a dummy load to the antenna
C. Put a "Danger - High Voltage" sign in the station
D. Put fuses in the main power line

A It is important to keep any unauthorized persons from using your Amateur Radio station. This is both a safety concern and a requirement of the FCC Rules. Disconnecting power and microphones from your equipment is an easy way to prevent unauthorized use of your station.

T2D10 **Why are unlicensed persons in your family not allowed to transmit on your amateur station if you are not there?**

A. They must not use your equipment without your permission
B. They must be licensed before they are allowed to be control operators
C. They must know how to use proper procedures and Q signals
D. They must know the right frequencies and emissions for transmitting

B [97.109(b)] - You may not allow an unlicensed person—even a family member—-to operate your radio transmitter while you are not present. They must first be licensed before they're allowed to be control operators.

T2D11 **When is it permissible for the control operator of a club station to accept compensation for sending information bulletins or Morse code practice?**

A. When compensation is paid from a non-profit organization
B. When the club station license is held by a non-profit organization
C. Anytime compensation is needed
D. When the station makes those transmissions for at least 40 hours per week

D [97.113(d)] - This is a special exception to the "no pecuniary interest" rules. The station may only be controlled by a paid operator when it is sending bulletins or Morse code practice. For example, the paid operator can not send bulletins and then engage in casual ragchews afterwards while being paid by the club.

Subelement T3

Good Operating Practices

Your Technician exam (element 2) will consist of 35 questions taken from the Technician question pool as prepared by the Volunteer Examiner Coordinator's Question Pool Committee. A certain number of questions are taken from each of the 10 subelements. There will be 4 questions from the subelement shown in this chapter. These questions are divided into 4 groups labeled T3A through T3D.

You'll occasionally see a reference to Part 97 of the Federal Communications Commission Rules set aside in brackets, like this: [97.3(a) (5)]. This tells you where to find the exact working of the Rules as they relate to that question. You'll find the complete Part 97 Rules on the ARRL Web site at **www.arrl.org/FandES/field/regulations/news/part97/**, and in the *FCC Rule Book* published by the ARRL.

T3A Choosing an operating frequency, calling CQ, calling another station, test transmissions – 1 exam question

T3A01 Which of the following should you do when selecting a frequency on which to transmit?

A. Call CQ to see if anyone is listening
B. Listen to determine if the frequency is busy
C. Transmit on a frequency that allows your signals to be heard
D. Check for maximum power output

B The first rule of good operating practice is to always listen before you transmit! This may seem so obvious that it's not worth a mention. However, a few seconds of listening will help ensure that you don't interfere with a contact (QSO) already in progress.

T3A02 How do you call another station on a repeater if you know the station's call sign?

A. Say "break, break" then say the station's call sign
B. Say the station's call sign then identify your own station
C. Say "CQ" three times then the other station's call sign
D. Wait for the station to call "CQ" then answer it

B To call another station when the repeater is not in use, just give both calls. For example, "N1IL, this is N1BKE." If the repeater is in use, but the conversation sounds like it is about to end, wait before calling another station.

T3A03 **How do you indicate you are looking for any station with which to make contact?**

A. CQ followed by your call sign
B. RST followed by your call sign
C. QST followed by your call sign
D. SK followed by your call sign

A CQ literally means "Calling any station." You can usually tell good operators by how they call CQ. A good operator makes short, crisp calls separated by listening periods. Think of a CQ as an advertisement for your station and your operating skills.

T3A04 **What should you transmit when responding to a call of CQ?**

A. Your own CQ followed by the other station's call sign
B. Your call sign followed by the other station's call sign
C. The other station's call sign followed by your call sign
D. A signal report followed by your call sign

C On phone, say the other station's call sign followed by "this is" and your call sign at least once using phonetics, such as those in Table T3-1. On CW, substitute the abbreviation DE for "this is". There is no need to send your call sign more than twice unless there is a lot of noise or interference present or the calling station is very weak.

Table T3-1
Standard ITU Phonetics

Letter	*Word*	*Pronunciation*	*Letter*	*Word*	*Pronunciation*
A	Alfa	**AL** FAH	N	November	NO **VEM** BER
B	Bravo	**BRAH** VOH	O	Oscar	**OSS** CAH
C	Charlie	**CHAR** LEE	P	Papa	PAH **PAH**
D	Delta	**DELL** TAH	Q	Quebec	KEH **BECK**
E	Echo	**ECK** OH	R	Romeo	**ROW** ME OH
F	Foxtrot	**FOKS** TROT	S	Sierra	SEE **AIR** RAH
G	Golf	GOLF	T	Tango	**TANG** GO
H	Hotel	HOH **TELL**	U	Uniform	**YOU** NEE FORM
I	India	**IN** DEE AH	V	Victor	**VIK** TAH
J	Juliett	**JEW** LEE ETT	W	Whiskey	**WISS** KEY
K	Kilo	**KEY** LOH	X	X-Ray	**ECKS** RAY
L	Lima	**LEE** MAH	Y	Yankee	**YANG** KEY
M	Mike	MIKE	Z	Zulu	**ZOO** LOO

Note: The **boldfaced** syllables are emphasized. The pronunciations shown in this table were designed for those who speak any of the international languages. The pronunciations given for "Oscar" and "Victor" may seem awkward to English-speaking people in the US.

T3A05 **What term describes a brief test transmission that does not include any station identification?**

A. A test emission with no identification required
B. An illegal un-modulated transmission
C. An illegal unidentified transmission
D. A non-voice ID transmission

C [97.119(a)] - No matter how brief or weak, a test transmission is not exempt from the rules requiring identification of the transmitting station. All identification rules must be followed. Test transmissions should be brief, but still must be identified. For example, instead of just pressing the push-to-talk switch on your handheld transceiver to see if your signal is strong enough to access a distant repeater, say your call sign.

T3A06 **What must an amateur do when making a transmission to test equipment or antennas?**

A. Properly identify the station
B. Make test transmissions only after 10:00 PM local time
C. Notify the FCC of the test transmission
D. State the purpose of the test during the test procedure

A See question T3A05

T3A07 **Which of the following is true when making a test transmission?**

A. Station identification is not required if the transmission is less than 15 seconds
B. Station identification is not required if the transmission is less than 1 watt
C. Station identification is required only if your station can be heard
D. Station identification is required at least every ten minutes and at the end of every transmission.

D See question T3A05

T3A08 **What is the meaning of the procedural signal "CQ"?**

A. Call on the quarter hour
B. New antenna is being tested (no station should answer)
C. Only the called station should transmit
D. Calling any station

D See question T3A03.

T3A09 **Why should you avoid using cute phrases or word combinations to identify your station?**

A. They are not easily understood by some operators
B. They might offend some operators
C. They do not meet FCC identification requirements
D. They might be interpreted as codes or ciphers intended to obscure your identification

A [97.119(b)(2)] - Whether you're working a DX operator who may not fully understand our language, or talking to your friend down the street, avoid using cute phrases or word combinations to identify your station. These can be confusing to anybody and they are not easily understood by non-English-speaking amateurs.

T3A10 **What brief statement is often used in place of "CQ" to indicate that you are listening for calls on a repeater?**

A. Say "Hello test" followed by your call sign
B. Say your call sign
C. Say the repeater call sign followed by your call sign
D. Say the letters "QSY" followed by your call sign

B Since a repeater's signal is generally strong and the output frequency fixed, there is no need for an extended CQ call as there is when using SSB or CW. All that's necessary is to announce your presence and anyone that wishes to call you will then do so. It is also common to say "monitoring" following your call sign to reinforce that you are present and listening.

T3A11 **Why should you use the International Telecommunication Union (ITU) phonetic alphabet when identifying your station?**

A. The words are internationally recognized substitutes for letters
B. There is no advantage
C. The words have been chosen to represent amateur radio terms
D. It preserves traditions begun in the early days of amateur radio

A [97.119(b)(2)] - If the other operator is having difficulty copying your signals you should use the standard International Telecommunications Union (ITU) phonetic alphabet, detailed in Table T3-1. Use the words in the phonetic alphabet to spell out the letters in your call sign, your name or any other piece of information that might be confused if the letters are not received correctly. This phonetic alphabet is generally understood by hams in all countries.

T3B Use of minimum power, band plans, repeater coordination, mode restricted sub-bands – 1 exam question

T3B01 What is a band plan?

A. A voluntary guideline, beyond the divisions established by the FCC for using different operating modes within an amateur band
B. A guideline from the FCC for making amateur frequency band allocations
C. A guideline for operating schedules within an amateur band published by the FCC
D. A plan devised by a local group

A Band plans are voluntary agreements between operators about how to use the bands. These plans go into more detail on using different operating modes within an amateur band than is specified in FCC regulations. Good operators are familiar with the band plans and try to follow them.

T3B02 Which of the following statements is true of band plans?

A. They are mandated by the FCC to regulate spectrum use
B. They are mandated by the ITU
C. They are voluntary guidelines for efficient use of the radio spectrum
D. They are mandatory only in the US

C Band plans are developed over time as amateurs determine the most efficient way to make use of the amateur bands within the constraints of the FCC Rules. Amateurs use them as voluntary guidelines perfectly suitable under most circumstances.

T3B03 Who developed the band plans used by amateur radio operators?

A. The US Congress
B. The FCC
C. The amateur community
D. The Interstate Commerce Commission

C Band plans are developed by entirely by the community of amateurs and amateur organizations.

T3B04 Who is in charge of the repeater frequency band plan in your local area?

A. The local FCC field office
B. RACES and FEMA
C. The recognized frequency coordination body
D. Repeater Council of America

C Volunteer frequency coordinators have been created to insure that new repeaters use frequencies that will tend not to interfere with existing repeaters in the same area. The FCC encourages frequency coordination, but the process is organized and run by hams and groups of ham who use repeaters.

T3B05 What is the main purpose of repeater coordination?

A. To reduce interference and promote proper use of spectrum
B. To coordinate as many repeaters as possible in a small area
C. To coordinate all possible frequencies available for repeater use
D. To promote and encourage use of simplex frequencies

A Having repeaters distributed randomly across the band is wasteful of spectrum and leads to interference between repeaters and other users. Having a means of coordinating the repeater input and output frequency assignments reduces interference and makes using repeaters easier for everyone.

T3B06 Who is accountable if a repeater station inadvertently retransmits communications that violate FCC rules?

A. The repeater trustee
B. The repeater control operator
C. The transmitting station
D. All of these answers are correct

C [97.205(g)] - Although there must be a control operator responsible for proper operation of the repeater, the primary responsibility for complying with FCC Rules for operating lies with the transmitting station. If repeater users consistently violate operating rules, the FCC can require that a repeater system be placed on remote control, meaning a control operator must monitor repeater operation.

T3B07 Which of these statements is true about legal power levels on the amateur bands?

A. Always use the maximum power allowed to ensure that you complete the contact
B. An amateur may use no more than 200 watts PEP to make an amateur contact
C. An amateur may use up to 1500 watts PEP on any amateur frequency
D. An amateur must use the minimum transmitter power necessary to carry out the desired communication

D [97.313(a)] - Although the maximum power level available to amateurs is 1500 watts PEP, the vast majority of contacts require far less power. Amateurs are required to avoid using excessive power levels to allow more hams to use the frequency. This does not mean you must reduce power until the other operator is barely able to hear you, just use a power level that provides satisfactory results.

T3B08 **Which of the bands available to Technician class licensees have mode restricted sub-bands?**

A. The 6-meter, 2-meter, and 70-centimeter bands
B. The 2-meter and 13-centimeter bands
C. The 6-meter, 2-meter, and 1¼-meter bands
D. The 2-meter and 70-centimeter bands

C [97.305(c)] - Referring to the table of operating privileges in the Introduction, you will see that above 50 MHz, the only bands that are divided in any way are 6-meters, 2-meters, and 1¼-meters. Both 6-meters and 2-meters have small CW-only sub-bands. 1¼- meters has a separated sub-band from 219-220 MHz in which data signals of less than 50 watts PEP are permitted. Otherwise, there are no mode restrictions in and above the 70-cm band. All of the separation of modes on those bands is according to band plans.

T3B09 **What emission modes are permitted in the restricted sub-band at 50.0-50.1 MHz?**

A. CW only
B. CW and RTTY
C. SSB only
D. CW and SSB

A [97.305 (a)(c)] - A small CW segment is set aside for beacon stations and types of operation that involve very weak signals, such as moonbounce and meteor scatter.

T3B10 **What emission modes are permitted in the restricted sub-band at 144.0-144.1 MHz?**

A. CW only
B. CW and RTTY
C. SSB only
D. CW and SSB

A [97.305 (a)(c)] - See question T3B09.

T3B11 ***This question has been withdrawn.***

T3C Courtesy and respect for others, sensitive subject areas, obscene and indecent language – 1 exam question

T3C01 What is the proper way to break into a conversation between two stations that are using the frequency?

A. Say your call sign between their transmissions
B. Wait for them to finish and then call CQ
C. Say "Break-break" between their transmissions
D. Call one of the operators on the telephone to interrupt the conversation

A When you wish to break into a conversation on a repeater, the proper procedure is to give your call sign as soon as one of the stations stops transmitting. Both stations should pause to listen for any stations breaking in (called *breakers*.) If you are one of the stations involved in an ongoing conversation, make sure you briefly pause before beginning your transmissions. This allows other stations to break in—there could be an emergency.

T3C02 What is considered to be proper repeater operating practice?

A. Monitor before transmitting and keep transmissions short
B. Identify legally
C. Use the minimum amount of transmitter power necessary
D. All of these answers are correct

D Most repeaters are not intended to support extended conversations, so keep contacts down to a few minutes or change to a simplex channel. A repeater is intended to provide communications over a wide area for lots of hams. Long conversations also run the risk of blocking emergency communications. As for all amateur communications, proper identification is required along with using only as much transmitter power as is needed.

T3C03 What should you do before responding to another station's call?

A. Make sure you are operating on a permissible frequency for your license class
B. Adjust your transmitter for maximum power output
C. Ask the station to send their signal report and location
D. Verify the other station's license class

A No matter what band you are using, until you upgrade to Amateur Extra it's a good idea to mentally check to be sure you're authorized to transmit before pressing the microphone push-to-talk switch!

T3C04 What rule applies if two amateur stations want to use the same frequency?

A. The station operator with a lesser class of license must yield the frequency to a higher-class licensee
B. The station operator with a lower power output must yield the frequency to the station with a higher power output
C. No frequency will be assigned for the exclusive use of any station and neither has priority
D. Station operators in ITU Regions 1 and 3 must yield the frequency to stations in ITU Region 2

C [97.101(b)] - If we are to make the best use of the limited amount of available spectrum, there must be ways to ensure that harmful interference is kept to a minimum. The FCC Rules require amateurs to cooperate with each other. If two amateur stations want to use the same frequency, both stations have an equal right to do so. This means that courtesy and common sense should be applied in order to avoid unnecessary conflict.

T3C05 Why is indecent and obscene language prohibited in the Amateur Service?

A. Because it is offensive to some individuals
B. Because young children may intercept amateur communications with readily available receiving equipment
C. Because such language is specifically prohibited by FCC Rules
D. All of these choices are correct

D [97.113(a)(4)] - Amateurs may not use obscene or indecent language—it is prohibited by the Rules. Amateur radio transmissions are public and anyone of any age can hear them. Depending on conditions and frequency, they might be heard anywhere in the world. There is no list of "words you can't say on Amateur Radio", but bear in mind the public nature of our communications and avoid any questionable language.

T3C06 Why should amateur radio operators avoid the use of racial or ethnic slurs when talking to other stations?

A. Such language is prohibited by the FCC
B. It is offensive to some people and reflects a poor public image on all amateur radio operators
C. Some of the terms used may be unfamiliar to other operators
D. You transmissions might be recorded for use in court

B Amateurs come from all walks of life and are distributed widely around the globe. With such a diverse population, it is hard to imagine a less desirable type of speech than racial and ethnic slang or slurs. Inconsiderate at best, such speech has no place in a public media such as Amateur Radio.

T3C07 What should you do if you hear a newly licensed operator that is having trouble with their station?

A. Tell them to get off the air until they learn how operate properly
B. Report them to the FCC
C. Contact them and offer to help with the problem
D. Move to another frequency

C One of the best traditions in Amateur Radio is that of mentoring newcomers to the hobby. Called "Elmering," more experienced operators serve as troubleshooters and references just as they were helped at the start of their ham careers. If you come across an operator having difficulty with a piece of equipment or a procedure, alert them to the problem with a helpful manner and offer to assist them. If necessary, contact them by telephone or email to prevent additional problems on the air.

T3C08 Where can an official list be found of prohibited obscene and indecent words that should not be used in amateur radio?

A. On the FCC web site
B. There is no official list of prohibited obscene and indecent words
C. On the Department of Commerce web site
D. The official list is in public domain and found in all amateur study guides

B [97.113(a)(4)] - See question T3C05.

T3C09 What type of subjects are not prohibited communications while using amateur radio?

A. Political discussions
B. Jokes and stories
C. Religious preferences
D. All of these answers are correct

D [97.113(a)(4)] - The FCC does not regulate the content of amateur communications except for indecent and obscene speech. Nevertheless, hams avoid topics that often lead to strong feelings, such as politics, religion, and those of a sexual nature. Bear in mind that telling a joke or story can be understood very differently than you intended because others can't see you to pick up on visual cues and mannerisms. Furthermore, you never know who is going to be listening. There are plenty of forums for these topics other than Amateur Radio.

T3C10 When circumstances are not specifically covered by FCC rules, what general operating standard must be applied to amateur station operation?

A. Designated operator control
B. Politically correct control
C. Good engineering and amateur practices
D. Reasonable operator control

C [97.101 (a)] - The FCC rules provide amateurs with a tremendous amount of flexibility to experiment and develop methods of communicating. After all, those are part of our service's basis and purpose for existence! In response, amateurs have developed a wide range of "good practices" for both technical and operating topics. Two good sources of information about good practices are *The ARRL Radio Handbook* and *The ARRL Operating Manual.*

T3D Interference to and from consumer devices, public relations, intentional and unintentional interference - 1 exam question

T3D01 What should you do if you receive a report that your transmissions are causing splatter or interference on nearby frequencies?

A. Increase transmit power
B. Change mode of transmission
C. Report the interference to the equipment manufacturer
D. Check transmitter for off frequency operation or spurious emissions

D Your equipment can cause spurious emissions if you operate it with some controls adjusted improperly. For example, if you set your microphone gain too high on an SSB transmitter, the resulting over-modulation creates spurious signals on nearby frequencies. On FM, misadjusting your tuning control or speaking too loudly (causing excessive deviation) can cause your signal to encroach on an adjacent channel.

T3D02 Who is responsible for taking care of the interference if signals from your transmitter are causing front end overload in your neighbor's television receiver?

A. You alone are responsible, since your transmitter is causing the problem
B. Both you and the owner of the television receiver share the responsibility
C. The FCC must decide if you or the owner of the television receiver is responsible
D. The owner of the television receiver is responsible

D You should realize that there is nothing you can do to your transmitter to cure receiver overload. It is a fundamental problem with the receiving system and the primary responsibility for curing the problem is with the owner. The receiver manufacturer should help, but it is the owner that must take the initiative.

T3D03 **What is the major cause of telephone interference?**

A. The telephone wiring is inadequate
B. Tropospheric ducting at UHF frequencies
C. The telephone was not equipped with adequate interference protection when manufactured.
D. Improper location of the telephone in the home

C Radio frequency energy from your amateur transmitter may interfere with your own or your neighbor's consumer electronic equipment, and that includes telephones. As with receiver overload, there is nothing you can do at the transmitter to cure the interference. Interference protection measures must be taken at, or in, the device in question. The major cause of telephone interference comes from telephones that were not equipped with interference protection when they were manufactured. (Reference: FCC CIB Telephone Interference Bulletin)

T3D04 **What is the proper course of action if you unintentionally interfere with another station?**

A. Rotate your antenna slightly
B. Properly identify your station and move to a different frequency
C. Increase power
D. Change antenna polarization

B Accidental interference happens frequently—perhaps you didn't hear the other stations before transmitting. Simple courtesy works wonders! Just say "excuse me," give your call sign, and move to a different frequency.

T3D05 **When may you deliberately interfere with another station's communications?**

A. Only if the station is operating illegally
B. Only if the station begins transmitting on a frequency you are using
C. Never
D. You may cause deliberate interference because it can't be helped during crowded band conditions

C [97.101(d)] - Although some interference is often unavoidable, deliberately operating in such a way as to interfere with another amateur's communications is called willful interference and is not permitted under any circumstances. For example, you must not repeatedly transmit on a frequency that is already occupied. Whether the frequency is occupied by a net or just two hams having a conversation, that would be willful interference.

T3D06 **Who has exclusive use of a specific frequency when the FCC has not declared a communication emergency?**

A. Any net station that has traffic
B. The station first occupying the frequency
C. Individuals passing health and welfare communications
D. No station has exclusive use of any frequency

D Unless the FCC has specifically indicated that a communications emergency is in effect, no amateur has more right to a frequency than any other, for any purpose. Amateurs are expected to be flexible in their operating plans and procedures so as to make best use of our spectrum.

T3D07 **What effect might a break in a cable television transmission line have on amateur communications?**

A. A break cannot affect amateur communications
B. Harmonic radiation from the TV may cause the amateur transmitter to transmit off-frequency
C. TV interference may result when the amateur station is transmitting, or interference may occur to the amateur receiver
D. The broken cable may pick up very high voltages when the amateur station is transmitting

C Any loose connector or break in the transmission line of a cable TV system can allow amateur signals to "leak" into the line, causing interference to TV receivers. Such a leak can also allow Cable TV signals to leak out of the system and cause interference to amateur receivers using that frequency. Cable TV systems typically use some amateur VHF/UHF frequencies to carry the signals, but only *inside* the cable. This causes no problems as long as there are no leaks in the system.

T3D08 **What is the best way to reduce on the air interference when testing your transmitter?**

A. Use a short indoor antenna when testing
B. Use upper side band when testing
C. Use a dummy load when testing
D. Use a simplex frequency instead of a repeater frequency

C The best way, of course, is to not put a signal on the air in the first place if it's not necessary! A dummy load (or dummy antenna) simulates an antenna while absorbing the transmitter output power and dissipating it as heat instead of radiating it as a signal. Most transmitter tests can be done using a dummy load instead of a live antenna, greatly reducing interference.

T3D09 **What rules apply to your station when using amateur radio at the request of public service officials or at the scene of an emergency?**

A. RACES
B. ARES
C. FCC
D. FEMA

C [97.103(a)] - Regardless of the situation, if you are operating an amateur station then you are required to comply with the FCC's Rules for the amateur service. If you are requested to use the station in a way that violates the FCC Rules (except in the case of there being an immediate threat to life or property), you should politely decline.

T3D10 **What do RACES and ARES have in common?**

A. They represent the two largest ham clubs in the United States
B. One handles road traffic, the other weather traffic
C. Neither may handle emergency traffic
D. Both organizations provide communications during emergencies

D RACES stands for Radio Amateur Civil Emergency Service, a communications service within the amateur service. RACES provides communications assistance to civil defense organizations in times of need. It is active only during periods of local, regional or national civil emergencies. ARES stands for Amateur Radio Emergency Service and is sponsored by the ARRL. ARES presents a way for local amateurs to provide emergency communications while working with local public safety agencies and groups such as the Red Cross. ARES can provide communications assistance at any time.

T3D11 **What is meant by receiver front-end overload?**

A. Too much voltage from the power supply
B. Too much current from the power supply
C. Interference caused by strong signals from a nearby source
D. Interference caused by turning the volume up too high

C Receiver overload is a common type of interference to TV and FM-broadcast receivers. It happens most often to consumer electronic equipment near an amateur station or other transmitter. When the RF signal (at the fundamental frequency) enters the receiver, it overloads one or more circuits. The receiver front end (first circuit stages after the antenna) is most commonly affected.

Subelement T4

Basic Electronics

Your Technician exam (element 2) will consist of 35 questions taken from the Technician question pool as prepared by the Volunteer Examiner Coordinator's Question Pool Committee. A certain number of questions are taken from each of the 10 subelements. There will be 5 questions from the subelement shown in this chapter. These questions are divided into 5 groups labeled T4A through T4E.

T4A Names of electrical units, DC and AC, what is a radio signal, conductors and insulators, electrical components - 1 exam question

T4A01 Electrical current is measured in which of the following units?

A. Volts
B. Watts
C. Ohms
D. Amperes

D The basic unit of electric current, a measure of the rate of flow of electrons, is the **ampere**, abbreviated **A**. It is named for Andre Marie Ampere, an early 19th century scientist who studied electricity extensively.

T4A02 Electrical Power is measured in which of the following units?

A. Volts
B. Watts
C. Ohms
D. Amperes

B The basic unit of electrical power is the **watt**, or **W**. This unit was named after James Watt, the 18th century inventor of the steam engine.

T4A03 **What is the name for the flow of electrons in an electric circuit?**

A. Voltage
B. Resistance
C. Capacitance
D. Current

D Electrons flow through the wires and components of an electric circuit. The flow of electrons in an electric circuit is called **current**.

T4A04 **What is the name of a current that flows only in one direction?**

A. An alternating current
B. A direct current
C. A normal current
D. A smooth current

B The electrons that make up **direct current**, or **dc**, flow in only one direction, travel from the negative battery terminal, through the circuit and back to the positive battery terminal.

T4A05 **What is the standard unit of frequency?**

A. The megacycle
B. The Hertz
C. One thousand cycles per second
D. The electromagnetic force

B The unit of measurement for frequency is Hertz, abbreviated Hz. A frequency of one cycle per second is one Hertz or 1 Hz. The unit is named for physicist Heinrich Hertz whose experiments in the late 19th century led to the development of radio.

T4A06 **How much voltage does an automobile battery usually supply?**

A. About 12 volts
B. About 30 volts
C. About 120 volts
D. About 240 volts

A Some batteries are actually made up of several internal batteries, called cells, connected in series to add the voltages from each cell. An automobile's 12-volt battery is made of six such cells, each supplying about 2 volts, for a total of 12 volts at the battery's external terminals.

T4A07 **What is the basic unit of resistance?**

A. The volt
B. The watt
C. The ampere
D. The ohm

D The **ohm** is the basic unit used to measure circuit resistance. It is named for Georg Simon Ohm, a German physicist and mathematician who discovered the relationship between voltage, current and resistance we call **Ohm's Law**.

T4A08 What is the name of a current that reverses direction on a regular basis?

A. An alternating current
B. A direct current
C. A circular current
D. A vertical current

A **Alternating current**, or **ac**, alternates direction, flowing first in one direction, then in the opposite direction.

T4A09 Which of the following is a good electrical conductor?

A. Glass
B. Wood
C. Copper
D. Rubber

C In general, most metals make good conductors of electricity because their electrons are relatively free to move in response to an applied voltage. Gold, silver and aluminum are all excellent conductors, while wood, paper and mica are very poor conductors.

T4A10 Which of the following is a good electrical insulator?

A. Copper
B. Glass
C. Aluminum
D. Mercury

B Most non-metallic materials act as insulators because their electrons are not free to move. That means very little current will flow in response to an applied voltage. Ceramics, glass, paper and mica are examples of good insulators.

T4A11 What is the term used to describe opposition to current flow in ordinary conductors such as wires?

A. Inductance
B. Resistance
C. Counter EMF
D. Magnetism

B Electrons flowing through any material collide with each other and with the individual atoms that make up the material. These collisions create an opposition to the flow of current called **resistance**.

T4A12 What instrument is used to measure the flow of current in an electrical circuit?

A. Frequency meter
B. SWR meter
C. Ammeter
D. Voltmeter

C A **meter** is a generic term for an instrument that displays a measured value as a number or as the position of a moving needle or pointer. Meters used to measure current, which is measured in amperes, are called **ammeters**.

T4A13 What instrument is used to measure Electromotive Force (EMF) between two points such as the poles of a battery?

A. Magnetometer
B. Voltmeter
C. Ammeter
D. Ohmmeter

B EMF is another term for **voltage**, which is measured in **volts**. A meter used to measure voltage is called a **voltmeter**.

T4B Relationship between frequency and wavelength, identification of bands, names of frequency ranges, types of waves – 1 exam question

T4B01 What is the name for the distance a radio wave travels during one complete cycle?

A. Wave speed
B. Waveform
C. Wavelength
D. Wave spread

C A radio wave travels at a constant speed, the speed of light. That means the radio wave will always cover the same distance during the time it takes the wave's oscillating electric and magnetic fields to make one complete cycle. That distance is the wave's **wavelength**. All radio waves of the same frequency, travelling at the speed of light, will have the same wavelength.

T4B02 What term describes the number of times that an alternating current flows back and forth per second?

A. Pulse rate
B. Speed
C. Wavelength
D. Frequency

D A complete sequence of ac current flowing, stopping, reversing, and stopping again is called a cycle. The number of cycles that occur each second is called **frequency**.

T4B03 What does 60 hertz (Hz) mean?

A. 6000 cycles per second
B. 60 cycles per second
C. 6000 meters per second
D. 60 meters per second

B One Hertz is the same as one cycle per second. (See also question T4A05.)

T4B04 Electromagnetic waves that oscillate more than 20,000 times per second as they travel through space are generally referred to as what?

A. Gravity waves
B. Sound waves
C. Radio waves
D. Gamma radiation

C The radio frequency spectrum is generally considered to begin at 20,000 Hz (20 kHz). Although radio communications can take place on frequencies lower than 20 kHz, this is the lowest practical frequency for RF use.

T4B05 How fast does a radio wave travel through space?

A. At the speed of light
B. At the speed of sound
C. Its speed is inversely proportional to its wavelength
D. Its speed increases as the frequency increases

A All electromagnetic energy—radio waves, light, X-rays--travels at the speed of light. In a vacuum, the speed of light (designated **c**) is 300,000,000 meters per second. It travels close to that speed in air. In denser materials, such as water, glass, and in cables, light travels slower.

T4B06 How does the wavelength of a radio wave relate to its frequency?

A. The wavelength gets longer as the frequency increases
B. The wavelength gets shorter as the frequency increases
C. There is no relationship between wavelength and frequency
D. The wavelength depends on the bandwidth of the signal

B Because radio waves travel at a constant speed, the higher their frequency, the less time it takes to complete one cycle, and the less distance it has traveled during that cycle—its wavelength. Higher frequencies mean shorter wavelengths and vice versa.

T4B07 What is the formula for converting frequency to wavelength in meters?

A. Wavelength in meters equals frequency in Hertz multiplied by 300
B. Wavelength in meters equals frequency in Hertz divided by 300
C. Wavelength in meters equals frequency in megahertz divided by 300
D. Wavelength in meters equals 300 divided by frequency in megahertz

D The key to remembering this relationship is that wavelength and frequency are **inversely proportional**. That is, as one gets larger, the other must get smaller. The only formula that satisfies this relationship is D: Wavelength (λ, in meters) = 300 / (Frequency (f) in MHz).

T4B08 What are sound waves in the range between 300 and 3000 Hertz called?

A. Test signals
B. Ultrasonic waves
C. Voice frequencies
D. Radio frequencies

C The human voice is generally composed of frequencies between 300 and 3000 Hz. There are some components of speech at higher and lower frequencies, but they are not critical to understanding speech. That is why most radios are designed to transmit just these **voice frequencies**.

T4B09 What property of a radio wave is often used to identify the different bands amateur radio operators use?

A. The physical length of the wave
B. The magnetic intensity of the wave
C. The time it takes for the wave to travel one mile
D. The voltage standing wave ratio of the wave

A Because of the radio wave's constant velocity (see question T4B01 and T4B06), knowing a radio wave's wavelength is the same as knowing its frequency and vice versa. By referring to a frequency band by approximate wavelength, which is traditional, the frequency range is also implied.

T4B10 What is the frequency range of the 2-meter band in the United States?

A. 144 to 148 MHz
B. 222 to 225 MHz
C. 420 to 450 MHz
D. 50 to 54 MHz

A You can use the formula f (in MHz) = 300 / Ω (in meters) to convert from wavelength to frequency. Remember that the wavelength designators for the ham bands (2-meters, 10-meters, etc.) are only approximate. Use the table of amateur bands in the Introduction to remember the exact frequency limits of each band.

T4B11 **What is the frequency range of the 6-meter band in the United States?**

A. 144 to 148 MHz
B. 222 to 225 MHz
C. 420 to 450 MHz
D. 50 to 54 MHz

D See question T4B10.

T4B12 **What is the frequency range of the 70-centimeter band in the United States?**

A. 144 to 148 MHz
B. 222 to 225 MHz
C. 420 to 450 MHz
D. 50 to 54 MHz

C See question T4B10.

T4C How radio works: receivers, transmitters, transceivers, amplifiers, power supplies, types of batteries, service life – 1 exam question

T4C01 **What is used to convert radio signals into sounds we can hear?**

A. Transmitter
B. Receiver
C. Microphone
D. Antenna

B A receiver is used to convert radio signals received by an antenna to audio signals. Most receivers have a speaker or headphones that convert the audio signals to sounds that we can hear. You can also connect the output of a receiver to a computer sound card or a data interface where the audio signals of a data signal are turned back into characters.

T4C02 **What is used to convert sounds from our voice into radio signals?**

A. Transmitter
B. Receiver
C. Speaker
D. Antenna

A Transmitters generate a steady radio signal and add information to it—our voice, Morse code from a key, or data from a computer. The information-carrying radio signal is then sent to an **antenna** through a **feed line**, where it is radiated as a radio wave that can be picked up with other antennas.

T4C03 What two devices are combined into one unit in a transceiver?

A. Receiver, transmitter
B. Receiver, transformer
C. Receiver, transistor
D. Transmitter, deceiver

A "Transceiver" is a combination of the words "transmitter" and "receiver." A **transmit-receive switch** is also required so that the transmitter and receiver sections can share a single antenna.

T4C04 What device is used to convert the alternating current from a wall outlet into low-voltage direct current?

A. Inverter
B. Compressor
C. Power Supply
D. Demodulator

C Most electronic equipment, including radios, consumes power as dc current supplied at a voltage of 30 V or less. Household electricity is supplied by the electric utility as ac at a voltage of 120 V. The power supply converts the ac input power to dc power suitable for the equipment.

T4C05 What device is used to increase the output of a 10-watt radio to 100 watts?

A. Amplifier
B. Power supply
C. Antenna
D. Attenuator

A Amplifiers are used to increase the strength of audio or radio signals. Amplifiers used to increase the power of transmitted signals are called **power amplifiers** or **linear amplifiers**. Amplifiers used to increase the strength of a received signal are called **pre-amplifiers** or just **pre-amps**.

T4C06 Which of the battery types listed below offers the longest life when used with a hand-held radio, assuming each battery is the same physical size?

A. Lead-acid
B. Alkaline
C. Nickel-cadmium
D. Lithium-ion

D Batteries based on Lithium-ion chemistry (the type of chemical reaction that supplies energy) have the highest **energy density**. This means that for batteries of equivalent size and weight, the Lithium-ion battery stores the most energy and so can keep a radio running the longest of the four types listed.

T4C07 **What is the nominal voltage per cell of a fully charged nickel-cadmium battery?**

A. 1.0 volts
B. 1.2 volts
C. 1.5 volts
D. 2.2 volts

B Each different type of battery chemistry supplies energy at a different **terminal voltage**. It is important to know what the terminal voltage of a battery is when selecting batteries for powering your radio equipment.

T4C08 **What battery type on this list is not designed to be re-charged?**

A. Nickel-cadmium
B. Carbon-zinc
C. Lead-acid
D. Lithium-ion

B The chemical reaction that supplies energy from a battery can not always be "run in reverse" to re-charge the battery. Attempting to re-charge a non-rechargeable battery often leads to overheating of the battery, corrosion, and possible damage to the re-charging equipment. Carbon-zinc and alkaline batteries are not re-chargeable.

T4C09 **What is required to keep rechargeable batteries in good condition and ready for emergencies?**

A. They must be inspected for physical damage and replaced if necessary
B. They should be stored in a cool and dry location
C. They must be given a maintenance recharge at least every 6 months
D. All of these answers are correct

D Proper maintenance is an important part of getting the most performance out of re-chargeable batteries. Keeping all batteries cool and dry slows the internal chemical reactions that tend to slowly discharge the battery's stored energy over time. Regular re-charging (called *cycling*) helps the battery store the maximum amount of energy.

T4C10 **What is the best way to get the most amount of energy from a battery?**

A. Draw current from the battery as rapidly as possible
B. Draw current from the battery at the slowest rate needed
C. Reverse the leads when the battery reaches the ½ charge level
D. Charge the battery as frequently as possible

B Slowly discharging a battery allows the chemical reaction that supplies energy to run at a slow rate, reducing the amount of heat and undesired by-products that build up as the chemicals are consumed. Overheating and by-products contaminating the original chemicals reduce the amount of energy that a battery can store.

T4D Ohms law relationships – 1 exam question

T4D01 What formula is used to calculate current in a circuit?

A. Current (I) equals voltage (E) multiplied by resistance (R)
B. Current (I) equals voltage (E) divided by resistance (R)
C. Current (I) equals voltage (E) added to resistance (R)
D. Current (I) equals voltage (E) minus resistance (R)

B Use the "Ohm's Law Circle" in Figure T4-1 by covering I, representing the quantity you don't know, leaving E (EMF or voltage) above R (resistance). Change that to the formula I = E / R.

Figure T4-1 — Ohm's Law Circle. If you know any two of voltage, current, or resistance, you can use Ohm's Law to find the third. In this figure, cover the missing quantity. If the remaining symbols are one over the other, divide the top quantity by the bottom quantity. If the remaining symbols are side-by-side, multiply those quantities.

T4D02 What formula is used to calculate voltage in a circuit?

A. Voltage (E) equals current (I) multiplied by resistance (R)
B. Voltage (E) equals current (I) divided by resistance (R)
C. Voltage (E) equals current (I) added to resistance (R)
D. Voltage (E) equals current (I) minus resistance (R)

A Use the "Ohm's Law Circle" in the figure by covering E, representing the quantity you don't know, leaving I (current) next to R (resistance). Change that to the formula E = I × R. Remember that Ohm's Law never involves addition or subtraction!

T4D03 What formula is used to calculate resistance in a circuit?

A. Resistance (R) equals voltage (E) multiplied by current (I)
B. Resistance (R) equals voltage (E) divided by current (I)
C. Resistance (R) equals voltage (E) added to current (I)
D. Resistance (R) equals voltage (E) minus current (I)

B Use the "Ohm's Law Circle" in the figure by covering R, representing the quantity you don't know, leaving E (EMF or voltage) above I (current). Change that to the formula R = E / I. Remember that Ohm's Law never involves addition or subtraction!

T4D04 What is the resistance of a circuit when a current of 3 amperes flows through a resistor connected to 90 volts?

A. 3 ohms
B. 30 ohms
C. 93 ohms
D. 270 ohms

B This is also an Ohm's Law question. Find the correct form of Ohm's Law by covering the quantity you don't know (in this case, R or resistance) in the figure. That leaves E (voltage) above I (current), so the formula is $R = E / I$.

$R = E / I$ $R = 90\ V / 3\ A = 30\ \Omega$ (ohms)

Notice that the equation symbol for EMF or voltage is E, but the units of voltage, volts, are abbreviated as V. Similarly, the symbol for current is I, but the units of current, amperes, are abbreviated as A. The equation symbol for resistance is R, but the Greek letter Ω is used to represent ohms.

T4D05 What is the resistance in a circuit where the applied voltage is 12 volts and the current flow is 1.5 amperes?

A. 18 ohms
B. 0.125 ohms
C. 8 ohms
D. 13.5 ohms

C Use the same procedure as in question T4D04. $R = 12\ V / 1.5\ A = 8\ \Omega$.

T4D06 What is the current flow in a circuit with an applied voltage of 120 volts and a resistance of 80 ohms?

A. 9600 amperes
B. 200 amperes
C. 0.667 amperes
D. 1.5 amperes

D Use the Ohm's Law Circle to find the formula for current; $I = E / R$. $I = 120\ V / 80\ \Omega = 1.5\ A$ or amperes.

T4D07 What is the voltage across the resistor if a current of 0.5 amperes flows through a 2-ohm resistor?

A. 1 volt
B. 0.25 volts
C. 2.5 volts
D. 1.5 volts

A Use the Ohm's Law Circle to find the formula for voltage; $E = I \times R$. $E = 0.5\ A \times 2\ \Omega = 1\ V$ or volt.

T4D08 **What is the voltage across the resistor if a current of 1 ampere flows through a 10-ohm resistor?**

A. 10 volts
B. 1 volt
C. 11 volts
D. 9 volts

A Use the Ohm's Law Circle to find the formula for voltage; $E = I \times R$. $E = 1\ A \times 10\ \Omega = 10\ V$ or volts.

T4D09 **What is the voltage across the resistor if a current of 2 amperes flows through a 10-ohm resistor?**

A. 20 volts
B. 0.2 volts
C. 12 volts
D. 8 volts

A Use the Ohm's Law Circle to find the formula for voltage; $E = I \times R$. $E = 2\ A \times 10\ \Omega = 20\ V$ or volts.

T4D10 **(C) What is the current flowing through a 100-ohm resistor connected across 200 volts?**

A. 20,000 amperes
B. 0.5 amperes
C. 2 amperes
D. 100 amperes

C Use the Ohm's Law Circle to find the formula for current; $I = E / R$. $I = 200\ V / 100\ \Omega = 2\ A$ or amperes.

T4D11 **What is the current flowing through a 24-ohm resistor connected across 240 volts?**

A. 24,000 amperes
B. 0.1 amperes
C. 10 amperes
D. 216 amperes

C Use the Ohm's Law Circle to find the formula for current; $I = E / R$. $I = 240\ V / 24\ \Omega = 10\ A$ or amperes.

T4E Power calculations, units, kilo, mega, milli, micro - 1 exam question

T4E01 What unit is used to describe electrical power?

A. Ohm
B. Farad
C. Volt
D. Watt

D See question T4A02.

T4E02 What is the formula used to calculate electrical power in a DC circuit?

A. Power (P) equals voltage (E) multiplied by current (I)
B. Power (P) equals voltage (E) divided by current (I)
C. Power (P) equals voltage (E) minus current (I)
D. Power (P) equals voltage (E) plus current (I)

A Power is the product of voltage and current. As with Ohm's Law, if you know any two of P, E, or I, you can determine the missing quantity from the Power Circle in Figure T4-2 as follows: $P = E \times I$ or $E = P / I$ or $I = P / E$. Voltage and current are never combined by adding or subtraction.

Figure T4-2 — Power Circle. Use the Power Circle as you use the Ohm's Law Circle by covering the missing quantity. Multiply or divide the remaining quantities based on whether they are side-by-side or one over the other.

T4E03 How much power is represented by a voltage of 13.8 volts DC and a current of 10 amperes?

A. 138 watts
B. 0.7 watts
C. 23.8 watts
D. 3.8 watts

A Use the Power Circle to find the formula for power, $P = E \times I$. $P = 13.8\ V \times 10\ A = 138\ W$ or watts.

T4E04 How much power is being used in a circuit when the voltage is 120 volts DC and the current is 2.5 amperes?

A. 1440 watts
B. 300 watts
C. 48 watts
D. 30 watts

B Use the Power Circle to find the formula for power, $P = E \times I$. $P = 120\ V \times 2.5\ A = 300\ W$ or watts.

T4E05 How can you determine how many watts are being drawn by your transceiver when you are transmitting?

A. Measure the dc voltage and divide it by 60 Hz
B. Check the fuse in the power leads to see what size it is
C. Look in the Radio Amateur's Handbook
D. Measure the dc voltage at the transceiver and multiply by the current drawn when you transmit

D This question assumes that your transceiver uses a dc power supply (most do.) Use the Power Circle to find the formula for power, P = E × I. That means you need to measure both voltage and current when you transmit.

T4E06 How many amperes are flowing in a circuit when the applied voltage is 120 volts DC and the load is 1200 watts?

A. 20 amperes
B. 10 amperes
C. 120 amperes
D. 5 amperes

B Use the Power Circle to find the formula for current, I = P / E. I = 1200 W / 120 V = 10 A or amperes.

T4E07 How many milliamperes is the same as 1.5 amperes?

A. 15 milliamperes
B. 150 milliamperes
C. 1500 milliamperes
D. 15000 milliamperes

C The prefix "milli" means "divide by 1000", so "milliamperes" means "divide by 1000 to get amperes." 1500 milliamperes / 1000 = 1.5 amperes.

T4E08 What is another way to specify the frequency of a radio signal that is oscillating at 1,500,000 Hertz?

A. 1500 kHz
B. 1500 MHz
C. 15 GHz
D. 150 kHz

A The prefix "kilo" means "multiply by 1000", so "kilohertz" means "multiply by 1000 to get hertz." 1500 kHz × 1000 = 1,500,000 hertz.

T4E09 How many volts are equal to one kilovolt?

A. one one-thousandth of a volt
B. one hundred volts
C. one thousand volts
D. one million volts

C The prefix "kilo" means "multiply by 1000", so "kilovolts" means "multiply by 1000 to get volts." 1 kilovolt × 1000 = 1000 volts.

T4E10 How many volts are equal to one microvolt?

A. one one-millionth of a volt
B. one million volts
C. one thousand kilovolts
D. one one-thousandth of a volt

A The prefix "micro" means "divide by 1,000,000", so "microvolts" means "divide by 1,000,000 to get volts." 1 microvolt / 1,000,000 = 0.000001 V or one-millionth of a volt.

T4E11 How many watts does a hand-held transceiver put out if the output power is 500 milliwatts?

A. 0.02 watts
B. 0.5 watts
C. 5 watts
D. 50 watts

B The prefix "milli" means "divide by 1000", so "milliwatts" means "divide by 1000 to get watts". 500 milliwatts / 1000 = 0.5 W or watts.

Subelement T5

Station Operation

Your Technician exam (element 2) will consist of 35 questions taken from the Technician question pool as prepared by the Volunteer Examiner Coordinator's Question Pool Committee. A certain number of questions are taken from each of the 10 subelements. There will be 4 questions from the subelement shown in this chapter. These questions are divided into 4 groups labeled T5A through T5D.

T5A Station hookup – microphone, speaker, headphones, filters, power source, connecting a computer – 1 exam question

T5A01 What does a microphone connect to in a basic amateur radio station?

A. The receiver
B. The transmitter
C. The SWR Bridge
D. The Balun

B Connected to the transmitter, a microphone converts the sound waves of your voice to electric current. The current, called an **audio signal**, is used by the transmitter to add your voice's information to the transmitted signal.

T5A02 Which piece of station equipment converts electrical signals to sound waves?

A. Frequency coordinator
B. Frequency discriminator
C. Speaker
D. Microphone

C Applied to a speaker, an audio signal causes a coil of wire to move back and forth. The coil is attached to a paper or plastic surface whose vibrations cause sound waves.

T5A03 What is the term used to describe what happens when a microphone and speaker are too close to each other?

A. Excessive wind noise
B. Audio feedback
C. Inverted signal patterns
D. Poor electrical grounding

B Audio feedback occurs when a speaker is reproducing the sound of the operator speaking into the microphone. If the microphone is close enough to the speaker or the speaker output is loud enough, the signal can build up to a loud squeal. This is called **feedback** because the speaker output is "fed back" into the microphone and transmitter. To prevent feedback, either configure your radio so that the speaker is off when you are transmitting, separate the microphone and speaker, or turn the speaker volume down.

T5A04 What could you use in place of a regular speaker to help you copy signals in a noisy area?

A. A video display
B. A low pass filter
C. A set of headphones
D. A boom microphone

C Using headphones blocks wind noise and noise from nearby people or machinery. Remember that the same noise may be picked up by your microphone, so take steps to shield it from noise, as well.

T5A05 What is a good reason for using a regulated power supply for communications equipment?

A. To protect equipment from voltage fluctuations
B. A regulated power supply has FCC approval
C. A fuse or circuit breaker regulates the power
D. Regulated supplies are less expensive

A To operate properly, communications equipment requires a power supply with a stable output voltage. Regulation prevents voltage fluctuations of the input power source from causing similar fluctuations at the output.

T5A06 Where must a filter be installed to reduce spurious emissions?

A. At the transmitter
B. At the receiver
C. At the station power supply
D. At the microphone

A Assuming the spurious emissions produced by the transmitter in question are "out of band," such as a harmonic, the only place that they can be filtered out before being transmitted is at the output of the transmitter.

T5A07 What type of filter should be connected to a TV receiver as the first step in trying to prevent RF overload from a nearby 2-meter transmitter?

A. Low-pass filter
B. High-pass filter
C. Band pass filter
D. Notch filter

D Because TV signals have frequencies both above and below the 2-meter band, a high-pass or low-pass filter will not do the job. Those filters would also attenuate some of the desired TV signals. Therefore, a notch filter that removes signals from a small range of frequencies--in this case the 2-meter ham band—must be used.

T5A08 What is connected between the transceiver and computer terminal in a packet radio station?

A. Transmatch
B. Mixer
C. Terminal Node Controller
D. Antenna

C The Terminal Node Controller (or **TNC**) converts the audio signals from your radio's receiver output to data characters for the computer. Similarly, characters from the computer are converted to audio signals that your transmitter can use. The TNC may also act as the push-to-talk switch to turn the transmitter on and off.

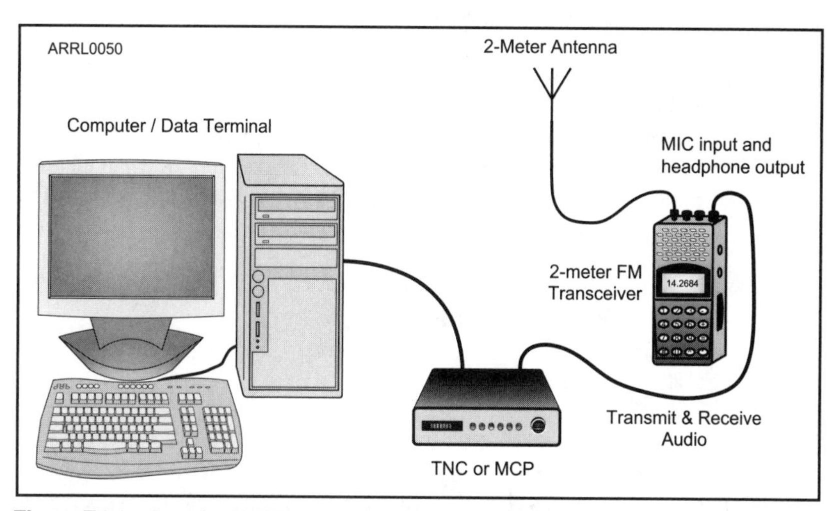

Figure T5-1 - A typical VHF packet station is composed of a computer or data terminal, a Terminal Node Controller (or TNC), and a VHF voice transceiver. The TNC converts between the data characters a computer can use and audio signals used by the transceiver.

T5A09 **Which of these items is not required for a packet radio station?**

A. Antenna
B. Transceiver
C. Power source
D. Microphone

D In a packet station, while audio is transmitted it is not voice. All communications is conducted as a stream of digital characters.

T5A10 **What can be used to connect a radio with a computer for data transmission?**

A. Balun
B. Sound Card
C. Impedance matcher
D. Autopatch

B Sound cards and their host computers have become sufficiently powerful that for several popular digital modes such as RTTY and PSK31, there is no need for a separate data interface. The sound card and software running on the host can perform the necessary conversions between audio signals and data characters.

T5B Operating controls – 1 exam question

T5B01 **What may happen if a transmitter is operated with the microphone gain set too high?**

A. The output power will be too high
B. It may cause the signal to become distorted and unreadable
C. The frequency will vary
D. The SWR will increase

B Your transmitter will produce spurious emissions if your microphone gain is too high. For example, on an SSB transmitter the resulting over-modulation creates spurious signals on nearby frequencies. On FM, the result will be excessive deviation that can cause your signal to encroach on an adjacent channel.

T5B02 **What kind of information may a VHF/UHF transceiver be capable of storing in memory?**

A. Transmit and receive operating frequency
B. CTCSS tone frequency
C. Transmit power level
D. All of these answers are correct

D By storing this information in memory, you can avoid having to reconfigure your transceiver when you change repeaters or simplex frequencies. Some radios can also store an alphanumeric label to make finding the desired frequency easier.

T5B03 What is one way to select a frequency on which to operate?

A. Use the keypad or VFO knob to enter the correct frequency
B. Turn on the CTCSS encoder
C. Adjust the power supply ripple frequency
D. All of these answers are correct

A The VFO (**Variable Frequency Oscillator**) is the circuit that controls the frequency of operation on both receive and transmit. Many radios also have a numeric keypad that allows you to enter the desired frequency directly. The VFO is used when you are tuning across the band looking for a station or particular signal. Keypads are used when changing directly between known frequencies.

T5B04 What is the purpose of the squelch control on a transceiver?

A. It is used to set the highest level of volume desired
B. It is used to set the transmitter power level
C. It is used to adjust the antenna polarization
D. It is used to quiet noise when no signal is being received

D In the absence of a signal on an FM receiver, you will hear noise. FM receivers have a **squelch** circuit that cuts off the speaker unless a signal is present, called "closing" the squelch. The squelch control adjusts the threshold at which squelch circuits turn the speaker on and off. The proper setting for the squelch control is just beyond the point at which the receiver audio is cut off. If set at a higher level, some signals will not be strong enough "open" the squelch.

T5B05 What is a way to enable quick access to a favorite frequency on your transceiver?

A. Enable the CTCSS tones
B. Store the frequency in a memory channel
C. Disable the CTCSS tones
D. Use the scan mode to select the desired frequency

B By storing the frequency and any other pertinent information in a memory channel you can quickly return to the frequency at any time. (See also question T5B02.)

T5B06 What might you do to improve the situation if the station you are listening to is hard to copy because of ignition noise interference?

A. Increase your transmitter power
B. Decrease the squelch setting
C. Turn on the noise blanker
D. Use the RIT control

C Ignition noise consists of a sharp pulse each time a spark plug in the engine fires. Just like the early spark radio transmitters, the arc produces RF signals across a wide range of frequencies. You hear the signals as buzzing or raspy noise that changes pitch along with engine speed. The noise blanker circuit in your radio detects these pulses and turns off the receiver for a short period.

T5B07 **What is the purpose of the buttons labeled "up" and "down" on many microphones?**

A. To allow easy frequency or memory selection
B. To raise or lower the internal antenna
C. To set the battery charge rate
D. To upload or download messages

A If the radio is set to use memory channels, the UP and DOWN controls select the memory channel to be used by the radio. If the radio is set to use VFO tuning (see question T5B03), the UP and DOWN controls tune the VFO one frequency step at a time.

T5B08 **What is the purpose of the "shift" control found on many VHF/UHF transceivers?**

A. Adjust transmitter power level
B. Change bands
C. Adjust the offset between transmit and receive frequency
D. Change modes

C Remember that a repeater listens on one frequency (where you transmit) and transmits on another (where you listen). The difference between those frequencies is the repeater's **offset** or **shift**. You must configure the radio to shift its transmit frequency to the repeater input frequency when you transmit and also be sure that the amount of shift is correct.

T5B09 **What does RIT mean?**

A. Receiver Input Tone
B. Receiver Incremental Tuning
C. Rectifier Inverter Test
D. Remote Input Transmitter

B Incremental Tuning is another way of saying "Fine Tuning". The RIT control allows you to adjust the receiver frequency a small amount while not affecting the transmitter frequency. RIT helps you tune in stations that are slightly off-frequency.

T5B10 **What is the purpose of the "step" menu function found on many transceivers?**

A. It adjusts the transmitter power output level
B. It adjusts the modulation level
C. It sets the earphone volume
D. It sets the tuning rate when changing frequencies

D A transceiver's VFO changes frequency in discrete steps. The STEP control (or menu item on some transceivers) allows you to change from small steps (for precise tuning adjustments) to large steps (for faster tuning). Some radios also have a FAST control that allows you to make a big change in frequency very quickly, but don't want to change the STEP size.

T5B11 **What is the purpose of the "function" or "F" key found on many transceivers?**

A. It turns the power on and off
B. It selects the autopatch access code
C. It selects an alternate action for some control buttons
D. It controls access to the memory scrambler

C Because front panel space is limited, many radios assign more than one function to specific controls. The function key is what changes the controls from one function to another.

T5C Repeaters; repeater and simplex operating techniques, offsets, selective squelch, open and closed repeaters, linked repeaters - 1 exam question

T5C01 **What is one purpose of a repeater?**

A. To cut your power bill by using someone else's higher power system
B. To extend the usable range of mobile and low-power stations
C. To transmit signals for observing propagation and reception
D. To communicate with stations in services other than amateur

B A repeater receives a signal and retransmits it, usually with higher power, better antennas and from a superior location, to provide an expanded communications range. VHF and UHF repeaters can greatly extend the operating range of amateurs using mobile and hand-held transceivers as shown in Figure T5.2. Instead of having to be within range of everyone you wish to talk to, you only have to be within range of the repeater.

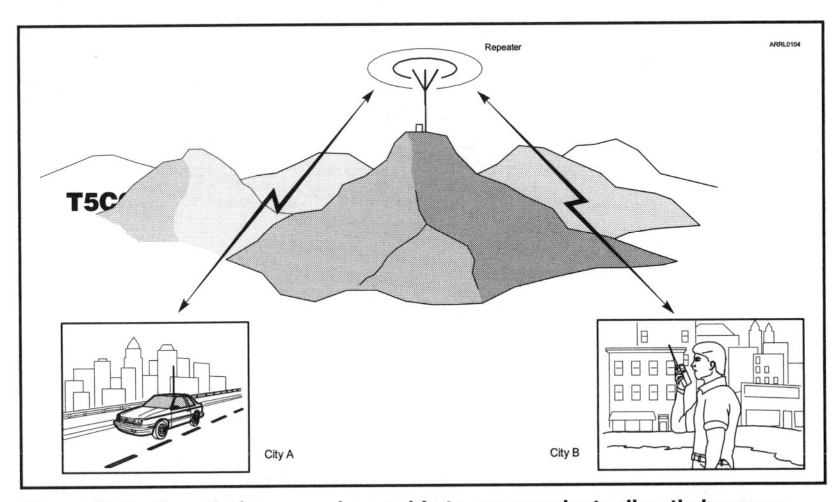

Figure T5-2 - Two stations may be unable to communicate directly because they are out of range or the signal path is blocked. However, a repeater located where both stations can access it can provide communications.

A. A tone used to identify the repeater
B. A tone used to indicate when a transmission is complete
C. A tone used to indicate that a message is waiting for someone
D. A tone used to activate a receiver in case of severe weather

B A courtesy tone sounds a second or so after the repeater receiver detects that the incoming transmission has ended. This prompts repeater users to leave some space between their transmissions so that other stations can break in. It also serves as a signal that the other station is really done transmitting and not just pausing. When the courtesy tone sounds, it is also a signal that the repeater's time-out timer has been reset and communications may continue without causing the repeater to shut down.

T5C03 Which of the following is the most important information to know before using a repeater?

A. The repeater input and output frequencies
B. The repeater call sign
C. The repeater power level
D. Whether or not the repeater has an autopatch

A To use a repeater you must be able to transmit and receive on the repeater's input and output frequencies, respectively. This means that you either have to know both frequencies or you have to know the repeater's output frequency and the required transmit offset.

T5C04 Why should you pause briefly between transmissions when using a repeater?

A. To let your radio cool off
B. To reach for pencil and paper so you can take notes
C. To listen for anyone wanting to break in
D. To dial up the repeater's autopatch

C If you are involved in an ongoing conversation, make sure you briefly pause before you begin each transmission. This allows other stations to break in—there could be an emergency. Don't key your microphone as soon as someone else stops transmitting. If you don't leave a short pause, others can't break in. This is why repeaters have time-out timers, to prevent users from transmitting continuously and preventing others from using the repeater.

T5C05 What is the most common input/output frequency offset for repeaters in the 2-meter band?

A. 0.6 MHz
B. 1.0 MHz
C. 1.6 MHz
D. 5.0 MHz

A On the 2-meter (144 to 148 MHz) band, most repeaters use an input/output frequency separation of 600 kHz. Sometimes the input frequency is higher than the output frequency (a *positive offset*) and sometimes lower (a *negative offset*). Table T5-1 summarizes the standard offsets on the various VHF and UHF bands.

Table T5-1

Repeater Input/Output Offsets

Band	*Offset*
6 meters	1 MHz
2 meters	600 kHz
1.25 meters	1.6 MHz
70 cm	5 MHz
33 cm	12 MHz
23 cm	20 MHz

T5C06 What is the most common input/output frequency offset for repeaters in the 70-centimeter band?

A. 600 kHz
B. 1.0 MHz
C. 1.6 MHz
D. 5.0 MHz

D See question T5C05. The offset on this band is larger because there is more spectrum in that band. By using a larger offset, there is less interference or interaction between the repeater's receiver and transmitter.

T5C07 What is meant by the terms input and output frequency when referring to repeater operations?

A. The repeater receives on one frequency and transmits on another
B. The repeater offers a choice of operating frequencies
C. One frequency is used to control the repeater and another is used to retransmit received signals
D. The repeater must receive an access code on one frequency before it will begin transmitting

A A repeater receives a signal on one frequency and simultaneously retransmits (repeats) it on another frequency. The frequency it receives on is called the **input frequency** and the frequency it transmits on is called the **output frequency**.

T5C08 What is the meaning of the term simplex operation?

A. Transmitting and receiving on the same frequency
B. Transmitting and receiving over a wide area
C. Transmitting on one frequency and receiving on another
D. Transmitting one-way communications

A In Amateur Radio, simplex operation means that the stations are communicating by transmitting and receiving on the same frequency. Using a repeater with different transmit and receive frequencies is called **duplex** operation.

T5C09 What is a reason to use simplex instead of a repeater?

A. When the most reliable communications are needed
B. To avoid tying up the repeater when direct contact is possible
C. When an emergency telephone call is needed
D. When you are traveling and need some local information

B Use simplex whenever you can make the contact without using a repeater. Repeaters must be shared over a wide area and by many amateurs. Use them wisely by not occupying the repeater unnecessarily.

T5C10 How might you find out if you could communicate with a station using simplex instead of a repeater?

A. Check the repeater input frequency to see if you can hear the other station
B. Check to see if you can hear the other station on a different frequency band
C. Check to see if you can hear a more distant repeater
D. Check to see if a third station can hear both of you

A The easiest way to determine if you are able to communicate with the other station on simplex is to listen to repeater's input frequency. Since this is the frequency the other station uses to transmit to the repeater, if you can hear the other station's signal, you should be able to use simplex.

T5C11 What is the term for a series of repeaters that can be connected to one another to provide users with a wider coverage?

A. Open repeater system
B. Closed repeater system
C. Linked repeater system
D. Locked repeater system

C By using radio links to relay the received signals from one repeater to another, it is possible to link repeaters together over very wide areas, such as an entire state or region. You can find out if a repeater is part of a linked system by contacting the repeater's control operator or sponsoring organization.

T5C12 **What is the main reason repeaters should be approved by the local frequency coordinator before being installed?**

A. Coordination minimizes interference between repeaters and makes the most efficient use of available frequencies
B. Coordination is required by the FCC
C. Repeater manufacturers have exclusive territories and you could be fined for using the wrong equipment
D. Only coordinated systems will be approved by the officers of the local radio club

A Frequency coordinators are able to analyze the location and power of proposed repeaters and determine whether there is the potential for interference to other repeaters in the region. By applying for coordination, a repeater owner can take advantage of the coordinator's expertise and avoid possible problems that could be difficult and expensive to solve. The coordinators are also very good at maximizing the use of the spectrum amateur's use.

T5C13 **Which of the following statements regarding use of repeaters is true?**

A. All amateur radio operators have the right to use any repeater at any time
B. Access to any repeater may be limited by the repeater owner
C. Closed repeaters must be opened at the request of any amateur wishing to use it
D. Open repeaters are required to use CTCSS tones for access

B Some repeaters have limited access and are called **closed repeaters**. The owner or owners may decide to restrict access to a small group, perhaps just the members of a club that maintain the repeater. Some repeaters are dedicated to specific functions, such as emergency communications. If you wish to join a group that sponsors such a repeater, contact the repeater's control operator or sponsoring club.

T5C14 **What term is used to describe a repeater when use is restricted to the members of a club or group?**

A. A beacon station
B. An open repeater
C. A auxiliary station
D. A closed repeater

D See question T5C13.

T5D Recognition and correction of problems, symptoms of overload and overdrive, distortion, over and under modulation, RF feedback, off frequency signals, fading and noise, problems with digital communications links – 1 exam question

T5D01 What is meant by fundamental overload in reference to a receiver?

A. Too much voltage from the power supply
B. Too much current from the power supply
C. Interference caused by very strong signals from a nearby source
D. Interference caused by turning the volume up too high

C Receiver overload is a common type of interference to TV and FM-broadcast receivers. It happens most often to consumer electronic equipment near an amateur station or other transmitter. When the RF signal (at the fundamental frequency) enters the receiver, it overloads one or more circuits. The receiver front end (first circuit stages after the antenna) is most commonly affected.

T5D02 Which of the following is NOT a cause of radio frequency interference?

A. Fundamental overload
B. Doppler shift
C. Spurious emissions
D. Harmonics

B Doppler shift refers to a change in frequency due to the transmitter and receiver moving towards or away from each other. It does not cause interference. All three of the other choices cause interference by disrupting receiver performance (A) or by creating unwanted signals that directly interfere with desired signals (C and D).

T5D03 What is the most likely cause of telephone interference from a nearby transmitter?

A. Harmonics from the transmitter
B. The transmitter's signals are causing the telephone to act like a radio receiver
C. Poor station grounding
D. Improper transmitter adjustment

B The major cause of telephone interference comes from telephones that were not equipped with interference protection when they were manufactured. (Reference: FCC CIB Telephone Interference Bulletin) Radio frequency energy from your amateur transmitter may be strong enough to cause the telephone to act as a receiver. As with receiver overload, there is nothing you can do at the transmitter to cure the interference. Interference protection measures must be taken at, or in, the telephone in question.

T5D04 **What is a logical first step when attempting to cure a radio frequency interference problem in a nearby telephone?**

A. Install a low-pass filter at the transmitter
B. Install a high-pass filter at the transmitter
C. Install an RF filter at the telephone
D. Improve station grounding

C If a telephone is responding to a strong nearby RF signal, the first step should be to prevent the RF from getting into the telephone in the first place. Since RF is probably entering the telephone through the wiring that connects it to the telephone network, an RF filter should be used. These filters are available at electronics stores. Cordless phones may be suffering from fundamental overload as described in question T5D03.

T5D05 **What should you do first if someone tells you that your transmissions are interfering with their TV reception?**

A. Make sure that your station is operating properly and that it does not cause interference to your own television
B. Immediately turn off your transmitter and contact the nearest FCC office for assistance
C. Tell them that your license gives you the right to transmit and nothing can be done to reduce the interference
D. Continue operating normally because your equipment cannot possibly cause any interference

A If a neighbor complains of television interference, you should first make sure that your equipment is operating properly and not creating spurious signals on the TV channels. Check for interference to your own TV. If you see it, stop operating and cure the problem before you go back on the air.

T5D07 **Which of the following may be useful in correcting a radio frequency interference problem?**

A. Snap-on ferrite chokes
B. Low-pass and high-pass filters
C. Notch and band-pass filters
D. All of these answers are correct

D Knowing how to use the different kinds of filters and RF blocking materials such as ferrite cores is an important part of the amateur's repertoire. If you are having an interference problem, members of your club may be able to help you. The ARRL also maintains an RF Interference resource Web page at **www.arrl.org/tis/info/rfigen.html**.

T5D08 What is the proper course of action to take when a neighbor reports that your radio signals are interfering with something in his home?

A. You are not required to do anything
B. Contact the FCC to see if other interference reports have been filed
C. Check your station and make sure it meets the standards of good amateur practice
D. Change your antenna polarization from vertical to horizontal

C While you may ultimately find out that the interference is the neighbor's responsibility, the first step you should always take is to check that your station is operating properly. The station construction and operation should always follow good amateur practices, as well.

T5D09 What should you do if a "Part 15" device in your neighbor's home is causing harmful interference to your amateur station?

A. Work with your neighbor to identify the offending device
B. Politely inform your neighbor about the rules that require him to stop using the device if it causes interference
C. Check your station and make sure it meets the standards of good amateur practice
D. All of these answers are correct

D Part 15 refers to the section of the FCC Rules that permits unlicensed devices to use radio frequencies as part of their function. This includes cordless phones, baby monitors, audio and video relay devices, and wireless computer data links. These devices are allowed to use the radio spectrum, but are not permitted to cause interference to stations in licensed services, such as Amateur Radio. This is one of the benefits of being a licensed station--protection against interference. The users of Part 15 devices are required to accept interference to the devices from stations in a licensed service, such as Amateur Radio or a broadcast station. These rules must be printed either directly on the device or in the owner's manual supplied with the device.

While this means that the responsibility for stopping the interference may lie with your neighbor, you should still make sure the interference is not occurring because of some problem in your station. If you are sure your station is operating properly, you may need to politely educate your neighbor about the interference and help identify the offending device.

T5D10 What could be happening if another operator tells you he is hearing a variable high-pitched whine on the signals from your mobile transmitter?

A. Your microphone is picking up noise from an open window
B. You have the volume on your receiver set too high
C. You need to adjust your squelch control
D. The power wiring for your radio is picking up noise from the vehicle's electrical system

D This phenomenon is called *alternator whine* and results from the electrical noise made by the alternator as it supplies power to the vehicle's electrical system. The tip-off is if the whine changes frequency with engine speed because the alternator speed is controlled by engine speed. A dc power filter, such as a large capacitor, at your radio's dc power input often reduces alternator whine.

T5D11 What may be the problem if another operator reports that your SSB signal is very garbled and breaks up?

A. You have the noise limiter turned on
B. The transmitter is too hot and needs to cool off
C. RF energy may be getting into the microphone circuit and causing feedback
D. You are operating on lower sideband

C If you can understand the other operator, you are probably not using the incorrect sideband. However, it is not uncommon for a microphone cable to pick up RF radiated by your antenna. This can disrupt the functioning of the transmitter's microphone input circuits. This is called *RF feedback* and usually results in distortion of the transmitted signal. Better grounding and ferrite cores often help reduce RF pickup by microphone input cabling.

T5D12 What might be the problem if you receive a report that your signal through the repeater is distorted or weak?

A. Your transmitter may be slightly off frequency
B. Your batteries may be running low
C. You could be in a bad location
D. All of these answers are correct

D Being slightly off frequency, such as if you have inadvertently mistuned your operating frequency by bumping a control knob or key, prevents the repeater receiver's detector from recovering your speech without distortion. Low batteries can also cause distortion if the transmitter's modulator circuits are unable to function properly and certainly the transmitter output circuits would not run at full power. Even being in a location where your signal can reach the repeater along multiple paths that interfere with each other can cause distortion. Moving about ½-wavelength at the frequency you're using may make a dramatic difference! With time, you will become experienced in recognizing these problems, preventing them from happening to you, and helping others to recognize and prevent them.

T5D13 What is one of the reasons to use digital signals instead of analog signals to communicate with another station?

A. Digital systems are less expensive than analog systems
B. Many digital systems can automatically correct errors caused by noise and interference
C. Digital modulation circuits are much less complicated than any other types
D. All digital signals allow higher transmit power levels

B The protocols used to exchange data by digital modes often contain methods for packaging the data along with specially coded information that helps the receiving equipment detect and sometimes correct errors. This can allow steady, reliable communications when signal levels are very low or in the presence of noise and fading.

Subelement T6

Communication Modes

Your Technician exam (element 2) will consist of 35 questions taken from the Technician question pool as prepared by the Volunteer Examiner Coordinator's Question Pool Committee. A certain number of questions are taken from each of the 10 subelements. There will be 3 questions from the subelement shown in this chapter. These questions are divided into 3 groups labeled T6A through T6C.

T6A Modulation modes, descriptions and bandwidth (AM, FM, SSB) – 1 exam question

T6A01 What are phone transmissions?

A. The use of telephones to set up an amateur radio contact
B. A phone patch between amateur radio and the telephone system
C. Voice transmissions by radio
D. Placing the telephone handset near a radio transceiver's microphone and speaker to relay a telephone call

C Any voice mode used for communications is known as a phone emission under FCC Rules. AM, SSB, and FM voice are all phone emission types. Phone is short for "radiotelephone," the full name for transmissions carrying voice information.

T6A02 Which of the following is a form of amplitude modulation?

A. Frequency modulation
B. Phase modulation
C. Single sideband
D. Phase shift keying

C An amplitude modulation (AM) signal consists of a carrier and two sidebands; one higher in frequency than the carrier (the upper sideband, USB) and one lower (the lower sideband, LSB). A single-sideband (SSB) signal is created by removing the carrier and one of the sidebands.

T6A03 What name is given to an amateur radio station that is used to connect other amateur stations to the Internet?

A. A gateway
B. A repeater
C. A digipeater
D. A beacon station

A Gateways are used by packet radio nodes and bulletin board systems to relay messages between the Internet and ham radio. Winlink, the system that allows amateurs to send and receive email via ham radio, uses gateways called PMBOs (Participating Mailbox Operators). Gateway stations are key components of new systems that combine Amateur Radio and the Internet.

T6A04 Which type of voice modulation is most often used for long distance and weak signal contacts on the VHF and UHF bands?

A. FM
B. AM
C. SSB
D. PM

C Single-sideband (SSB) is more effective for communicating with weak signals because the signal's power is contained in a small bandwidth compared to AM and FM/PM. SSB reception is affected by noise more than FM/PM, however.

T6A05 Which type of modulation is most commonly used for VHF and UHF voice repeaters?

A. AM
B. SSB
C. PSK
D. FM

D Frequency modulation (FM) and its close relative phase modulation (PM) are used in repeater communication because of the mode's superior rejection of noise and static.

T6A06 Which emission type has the narrowest bandwidth?

A. FM voice
B. SSB voice
C. CW
D. Slow-scan TV

C CW signals consist of a continuous wave signal on a single frequency that is turned on and off in the patterns that make up the Morse code. As a result, CW has the narrowest bandwidth of all modes used by amateurs.

T6A07 Which sideband is normally used for VHF and UHF SSB communications?

A. Upper sideband
B. Lower sideband
C. Suppressed sideband
D. Inverted sideband

A Normally, hams use upper sideband (USB) for SSB communications on VHF/UHF frequencies. This choice is by convention and not for technical reasons.

T6A08 What is the primary advantage of single sideband over FM for voice transmissions?

A. SSB signals are easier to tune in than FM signals
B. SSB signals are less likely to be bothered by noise interference than FM signals.
C. SSB signals use much less bandwidth than FM signals
D. SSB signals have no advantages at all in comparison to other modes.

C Whereas an FM signal may have a bandwidth of 5 to 15 kHz, SSB signals only require 2 to 3 kHz of spectrum. This increases the usable range of SSB signals compared to FM signals of equal power.

T6A09 What is the approximate bandwidth of a single-sideband voice signal?

A. 1 kHz
B. 2 kHz
C. Between 3 and 6 kHz
D. Between 2 and 3 kHz

D Most SSB signals contains voice frequencies from 300 to around 3000 Hz for a bandwidth of about 2.7 kHz. Under crowded or noisy conditions, amateurs can use filters in their transmitters to reduce signal bandwidth to as low as 2 kHz, trading fidelity for improved reception.

T6A10 What is the approximate bandwidth of a frequency-modulated voice signal?

A. Less than 500 Hz
B. About 150 kHz
C. Between 5 and 15 kHz
D. More than 30 kHz

C Frequency modulation (FM) and phase modulation (PM) signals achieve a good balance of voice fidelity and range for signal bandwidths of 5 to 15 kHz. Wider bandwidths are used by commercial FM broadcast stations to transmit programs with high fidelity, but much higher power is needed to achieve a useful reception range.

T6A11 What is the normal bandwidth required for a conventional fast-scan TV transmission using combined video and audio on the 70-centimeter band?

A. More than 10 MHz
B. About 6 MHz
C. About 3 MHz
D. About 1 MHz

B Because there is so much information in a fast-scan video signal, more bandwidth is required for its transmission. Fast-scan video (or ATV) is composed of an AM video signal that consumes most of the bandwidth and an FM sound signal.

T6B Voice communications, EchoLink and IRLP 1 exam question

T6B01 How is information transmitted between stations using EchoLink?

A. APRS
B. PSK31
C. Internet
D. Atmospheric ducting

C EchoLink uses Voice Over Internet Protocol (VoIP) technology to relay digitized voice information between repeaters and computers around the world. Users can connect to the EchoLink system via radio to a repeater or via the Internet by using a computer.

T6B02 **What does the abbreviation IRLP mean?**

A. Internet Radio Linking Project
B. Internet Relay Language Protocol
C. International Repeater Linking Project
D. International Radio Linking Project

A The IRLP system is a method of linking repeaters using Voice Over Internet Protocol (VoIP) technology. Hams connect to the IRLP system via a repeater and use control tones to connect to other repeaters on the system.

T6B03 **Who may operate on the EchoLink system?**

A. Only club stations
B. Any licensed amateur radio operator
C. Technician class licensed amateur radio operators only
D. Any person, licensed or not, who is registered with the EchoLink system

B Before being allowed to use the EchoLink system, you must send a copy of your amateur license to the EchoLink system administrators. Novice class licensees may only listen on EchoLink because they do not have privileges for the frequencies on which most EchoLink repeaters operate.

T6B04 **What technology do EchoLink and IRLP have in common?**

A. Voice over Internet protocol
B. Ionospheric propagation
C. AC power lines
D. PSK31

A Both use Voice Over Internet Protocol (VoIP) technology developed for computer-to-computer use on the Internet. VoIP is also used for Internet-based telephone service.

T6B05 **What method is used to transfer data by IRLP?**

A. VHF Packet radio
B. PSK31
C. Voice over Internet protocol
D. None of these answers are correct

C See questions T6B02 and T6B04.

T6B06 **What does the term IRLP describe?**

A. A method of encrypting data
B. A method of linking between two or more amateur stations using the Internet
C. A low powered radio using infrared frequencies
D. An international logging program.

B See questions T6B02 and T6B04.

T6B07 **Which one of the following allows computer-to-radio linking for voice transmission?**

A. Grid modulation
B. EchoLink
C. AMTOR
D. Multiplex

B Of the repeater linking systems in existence as of early 2006, EchoLink is the only one that allows a user to connect to and use the system from a computer.

T6B08 **What are you listening to if you hear a brief tone and then a station from Russia calling CQ on a 2-meter repeater?**

A. An ionospheric band opening on VHF
B. A prohibited transmission
C. An Internet linked DX station
D. None of these answers are correct

C When a connection between two repeaters is made, a special tone or message is transmitted on the repeater to which the connection is made. This alerts listeners that the repeater link has been established. The station initiating the connection, in this case from Russia, can then begin transmitting on the linked repeater.

T6B10 **Where might you find a list of active nodes using VoIP?**

A. The FCC Rulebook
B. From your local emergency coordinator
C. A repeater directory or the Internet
D. The local repeater frequency coordinator

C Both the IRLP and EchoLink systems maintain a directory of repeaters participating in their systems. You can browse the directories to find nodes (repeaters and Internet voice server systems) currently on the air.

T6B11 **When using a portable transceiver how do you select a specific IRLP node?**

A. Choose a specific CTCSS tone
B. Choose the correct DSC tone
C. Access the repeater autopatch
D. Use the keypad to transmit the IRLP node numbers

D Each repeater and node is identified by a four-digit numeric code. After the linking process has been initiated, the link to a specific node is made by transmitting the code for that node. The code then directs the repeater controller to establish the Internet link to the desired node.

T6C Non-voice communications - image communications, data, CW, packet, PSK31, Morse code techniques, Q signals – 1 exam question

T6C01 **Which of the following is an example of a digital communications method?**

A. Single sideband voice
B. Amateur television
C. FM voice
D. Packet radio

D Packet radio uses a version of the computer-to-computer X.25 data exchange protocol that was adapted to suit amateur use becoming the AX.25 protocol. In packet communications, data characters from a terminal or computer are packaged along with routing and error-detection information into a packet that is transmitted over the air.

T6C02 **What does the term APRS mean?**

A. Automatic Position Reporting System
B. Associated Public Radio Station
C. Auto Planning Radio Set-up
D. Advanced Polar Radio System

A The Automatic Position Reporting System (APRS) was created by WB4APR as a way of transmitting GPS location information over Amateur Radio. When using APRS, the location of an individual amateur station can be viewed on maps accessed via the Internet.

T6C03 What item is required along with your normal radio for sending automatic location reports?

A. A connection to the vehicle speedometer
B. A connection to a WWV receiver
C. A connection to a broadcast FM sub-carrier receiver
D. A Global Positioning System receiver

D Data from a GPS receiver is fed to a packet radio TNC and a VHF radio as shown in Figure T6-1. The packets are then transmitted to a local APRS node directly or via a local APRS digipeater.

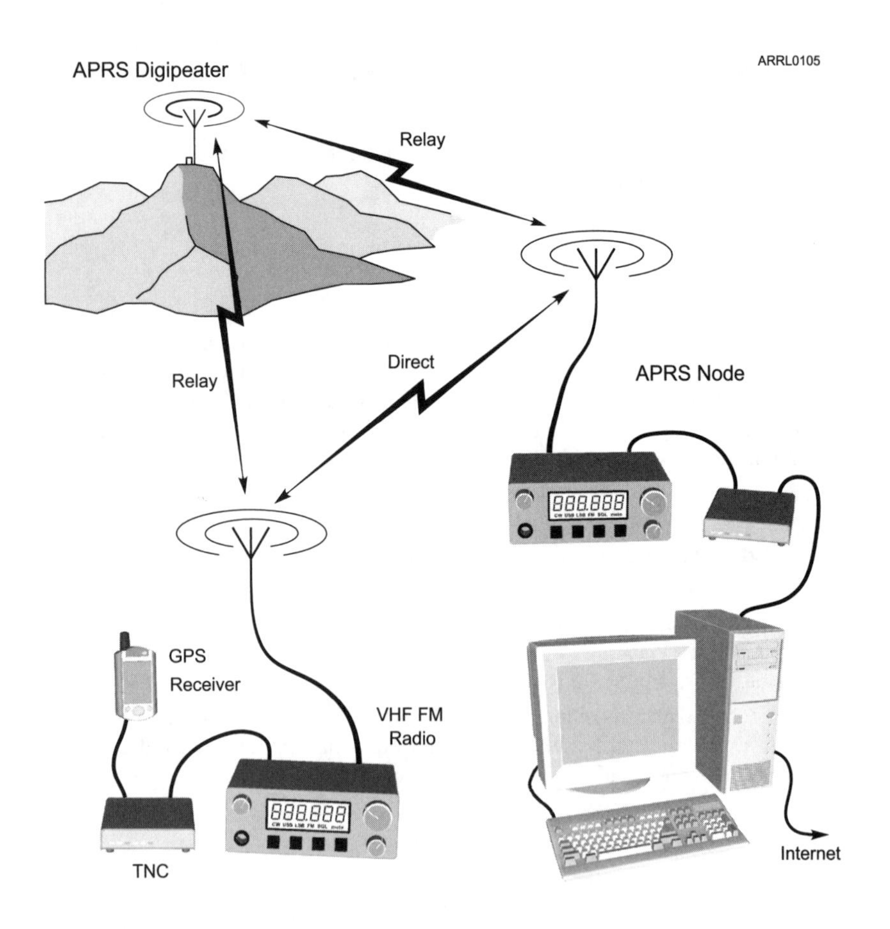

Figure T6.1 - To use the Automatic Position Reporting System (APRS), the output of a GPS receiver is connected to a packet radio TNC and a VHF radio. Position information and the call sign of the reporting station are then transferred to the ARPS system by an APRS node, either directly or via a digipeater. Station location can then be viewed via the Internet.

T6C04 **What type of transmission is indicated by the term NTSC?**

A. A Normal Transmission mode in Static Circuit
B. A special mode for earth satellite uplink
C. A standard fast-scan color television signal
D. A frame compression scheme for TV signal

C NTSC stands for the National Television System Committee, which developed the standards that define, broadcast fast-scan television signals. Amateurs adopted the NTSC standard for Amateur Television (ATV) transmissions. RS-170 is the standard that describes fast-scan video signals before they are sent over the air.

T6C05 **What emission mode may be used by a Technician class operator in the 219 - 220 MHz frequency range?**

A. Slow-scan television
B. Point-to-point digital message forwarding
C. FM voice
D. Fast-scan television

B A Technician class operator is allowed to use point-to-point digital message forwarding between 219 MHz and 220 MHz. Amateurs have access to these frequencies as a secondary allocation and are limited to 50 W PEP. Other emission types are not allowed in this part of the 1¼-meter band. Check the FCC rules [97.305 (c)] for other limitations before you begin operations.

T6C06 **What does the abbreviation PSK mean?**

A. Pulse Shift Keying
B. Phase Shift Keying
C. Packet Short Keying
D. Phased Slide Keying

B In Phase-Shift Keying, the phase of a signal is varied in order to convey information. Amateurs use PSK of an audio signal to transmit information. The audio signal can be transmitted as an AM or FM signal.

T6C07 What is PSK31?

A. A high-rate data transmission mode used to transmit files
B. A method of reducing noise interference to FM signals
C. A type of television signal
D. A low-rate data transmission mode that works well in noisy conditions

D The popular PSK31 modulation system uses PSK to send low-rate digital data. It was specially designed to work well in the presence of noise and fading that are common on the HF bands. As a result, it enables communications with very low power.

T6C08 What sending speed is recommended when using Morse code?

A. Only speeds below five WPM
B. The highest speed your keyer will operate
C. Any speed at which you can reliably receive
D. The highest speed at which you can control the keyer

C A station responding to what you send will probably send at about the same speed you do. If you send faster than you can receive, you won't be able to copy what the other station sends. It's also a good idea not to send so fast that you can't form the code characters properly, even if you can receive properly.

T6C09 What is a practical reason for being able to copy CW when using repeaters?

A. To send and receive messages others cannot overhear
B. To conform with FCC licensing requirements
C. To decode packet radio transmissions
D. To recognize a repeater ID sent in Morse code

D Repeater systems often identify themselves by sending their call sign in Morse code. If you are trying to determine who is the control operator or where the repeater is located, you'll need the call sign to look up the information.

T6C10 **What is the "Q" signal used to indicate that you are receiving interference from other stations?**

A. QRM
B. QRN
C. QTH
D. QSB

A Q-signals are a system of making queries and exchanging information in an abbreviated form. They also allow operators that speak different languages to communicate. QRM refers to interference from other stations. QRN refers to interference from atmospheric static, QTH means the station's location. QSB indicates signal fading. There are many useful Q-signals used by amateurs.

T6C11 **What is the "Q" signal used to indicate that you are changing frequency?**

A. QRU
B. QSY
C. QSL
D. QRZ

B QSY means, "I am changing frequency". (See also question T6C10.)

Subelement **T7**

Special Operations

Your Technician exam (element 2) will consist of 35 questions taken from the Technician question pool as prepared by the Volunteer Examiner Coordinator's Question Pool Committee. A certain number of questions are taken from each of the 10 subelements. There will be 2 questions from the subelement shown in this chapter. These questions are divided into 2 groups labeled T7A and T7B.

T7A Operating in the field, radio direction finding, radio control, contests, special event stations – 1 exam question

T7A01 What is a good thing to have when operating a hand-held transceiver away from home?

A. A selection of spare parts
B. A programming cable to load new channels
C. One or more fully charged spare battery packs
D. A dummy load

C Maintaining your charged battery packs in good condition and keeping them fully charged is not difficult, but does require regular attention. Sometimes it seems like batteries know when you're away from home and discharge faster on purpose! If you operate away from home frequently, have enough battery packs to provide power for the time you expect to spend operating.

T7A02 Which of these items would probably not be very useful to include in an emergency response kit?

A. An external antenna and several feet of connecting cable
B. A 1500-watt output linear amplifier
C. A cable and clips for connecting your transceiver to an external battery
D. A listing of repeater frequencies and nets in your area

B While an amplifier will certainly give you a strong signal, it requires large amounts of power, heavy duty feed line, antennas that can handle the high power, and safety measures to keep people away from the antenna. Most high-power amplifiers are heavy and difficult to move. All of these are reasons why amplifiers are best used at fixed stations and not for mobile or portable emergency response.

T7A03 **How can you make the signal from a hand-held radio stronger when operating in the field?**

A. Switch to VFO mode
B. Use an external antenna instead of the rubber-duck antenna
C. Stand so there is a metal building between you and other stations
D. Speak as loudly as you can

B The flexible "rubber-duck" antennas supplied with handheld radios are very convenient, but are not the most efficient antenna. You can increase your signal strength considerably by using a full-size antenna, such as a dipole or quarter-wave ground plane.

T7A04 **What would be a good thing to have when operating from a location that includes lots of crowd noise?**

A. A portable bullhorn
B. An encrypted radio
C. A combination headset and microphone
D. A pulse noise blanker

C Noise of any sort—crowd, machinery, wind—can make it difficult for you to hear the audio from your radio's speaker and at the same time make your speech hard to understand by other operators. Using headphones does wonders to improve intelligibility and fidelity of received audio. If the headphone also includes a microphone (sometimes called a *boom mike* or *boomset*), your transmitted audio will also improve because the microphone is kept near your mouth. As an added bonus, you don't have to hold the microphone or radio in one hand!

T7A05 **What is a method used to locate sources of noise interference or jamming?**

A. Echolocation
B. Doppler radar
C. Radio direction finding
D. Phase locking

C There are a number of techniques that allow a radio operator to determine the bearing to a radio transmitter. Using them is called **direction finding**. Exercises in which hams try to find a hidden transmitter are called *foxhunts* or *bunny hunts*.

T7A06 **Which of these items would be the most useful for a hidden transmitter hunt?**

A. Binoculars and a compass
B. A directional antenna
C. A calibrated noise bridge
D. Calibrated SWR meter

B The point of the exercise is to determine direction to a radio source, the hidden transmitter, and to do so requires a directional antenna.

T7A07 What is a popular operating activity that involves contacting as many stations as possible during a specified period?

A. Contesting
B. Net operations
C. Public service events
D. Simulated emergency exercises

A A radio contest consists of making as many contacts as you can with stations in a targeted area or on specific bands. Contests last for only a few hours or maybe all weekend. They're a great way to improve your station capabilities and operating skills.

T7A09 What is a grid locator?

A. A letter-number designator assigned to a geographic location
B. Your azimuth and elevation
C. Your UTC location
D. The 4 digits that follow your ZIP code

A The Maidenhead Locator System (named after the city in England where it was developed) divides the Earth's surface into grids organized by latitude and longitude. The designator for each **grid square** consists of two letter and two numbers, such as CN87 or FM13.

T7A10 What is a special event station?

A. A station that sends out birthday greetings
B. A station that operates only on holidays
C. A temporary station that operates in conjunction with an activity of special significance
D. A station that broadcasts special events

C Special event stations operate in conjunction with some noteworthy event, such as a festival, convention, or expedition to an unusual location. The station operates for a day to a couple of weeks and may also use a special call sign.

T7A11 What is the maximum power allowed when transmitting telecommand signals to radio controlled models?

A. 500 milliwatts
B. 1 watt
C. 25 watts
D. 1500 watts

B [97.215(c)] - Radio control takes place over relatively short distances, so there is no need for high power transmitters.

T7A12 What is the station identification requirement when sending commands to a radio control model using amateur frequencies?

A. Voice identification must be transmitted every 10 minutes
B. Morse code ID must be sent once per hour
C. A label indicating the licensee's call sign and address must be affixed to the transmitter
D. There is no station identification requirement for this service

C [97.215(a)] - There is not much point in sending a station call sign to a radio-controlled model. Because radio control signals are low power (see question T7A11), to identify the signal's source it is sufficient to put the call sign of the operator in control of the signal on the transmitter.

T7B Satellite operation, Doppler shift, satellite sub bands, LEO, orbit calculation, split frequency operation, operating protocols, AMSAT, ISS communications – 1 exam question

T7B01 What class of license is required to use amateur satellites?

A. Only Extra class licensees can use amateur radio satellites
B. General or higher-class licensees who have a satellite operator certification
C. Only persons who are AMSAT members and who have paid their dues
D. Any amateur whose license allows them to transmit on the satellite uplink frequency

D [97.209 (a)] - The control operator of a station communicating through an amateur satellite may hold any class of amateur license that has privileges to transmit in the satellite's input frequency range.

T7B02 How much power should you use to transmit when using an amateur satellite?

A. The maximum power of your transmitter
B. The minimum amount of power needed to complete the contact
C. No more than half the rating of your linear amplifier
D. Never more than 1 watt

B Using minimum power is particularly important when using satellites because of their very limited power sources. By transmitting with excessive power, you can actually cause the signals of other stations using the satellite to be reduced as they are relayed by the satellite.

T7B03 What is something you can do when using an amateur radio satellite?

A. Listen to the Space Shuttle
B. Get global positioning information
C. Make autopatch calls
D. Talk to amateur radio operators in other countries

D Because the satellite is high above the Earth, its transmitted signal can often be heard in many countries at the same time. The satellites farthest from Earth can provide coverage over nearly half of the Earth's surface!

T7B04 Who may make contact with an astronaut on the International Space Station using amateur radio frequencies?

A. Only members of amateur radio clubs at NASA facilities
B. Any amateur with a Technician or higher class license
C. Only the astronaut's family members who are hams
D. You cannot talk to the ISS on amateur radio frequencies

B The astronaut-amateurs on board the ISS operate according to the same FCC Rules that earthbound hams do, so there is no reason that communications between them should be restricted based on license class. Any amateur with a license that permits communications on the VHF frequencies used by the ISS is welcome to make contact!

T7B05 What is a satellite beacon?

A. The primary transmit antenna on the satellite
B. An indicator light that that shows where to point your antenna
C. A reflective surface on the satellite
D. A signal that contains information about a satellite

D Like its terrestrial counterparts, a satellite's beacon transmits communications so that propagation to the satellite and information about the satellite can be observed. Receiving the beacon from a satellite indicates that it is within range and contacts with it or through it can be attempted.

T7B06 What should you use to determine when you can access an amateur satellite?

A. A GPS receiver
B. A field strength meter
C. A telescope
D. A satellite-tracking program

D Satellites are typically only visible for short periods at a time, called a *pass*, because they are moving so rapidly. A computer can perform the calculations required to tell when a particular satellite is visible from your location by using a **satellite-tracking** program. The program uses mathematical coordinates called *Keplerian elements* available from on-line Web sites. The output of the program tells you when and where to look in the sky for the satellite.

T7B07 What is Doppler shift?

A. A change in the satellite orbit
B. A mode where the satellite receives signals on one band and transmits on another
C. A change in signal frequency caused by motion through space
D. A special digital communications mode for some satellites

C Doppler shift or Doppler effect describes the way the downlink frequency of a satellite varies by several kilohertz during an orbit due to its motion relative to you. As the satellite is moving toward you, the frequency of the satellite's signal will increase by a small amount. After the satellite passes overhead, the frequency of the signal will begin to decrease. This is the same effect that causes the sound of a vehicle's horn or whistle to change pitch as it moves past you.

T7B08 What is the name of the group that coordinates the building and/or launch of the largest number of amateur radio satellites?

A. NSA
B. USOC
C. AMSAT
D. FCC

C The Amateur Satellite Corporation (AMSAT) is an independent, non-profit company that is responsible for managing many of the amateur satellites. Amateur organizations in other countries also build and launch satellites of their own.

T7B09 What is a satellite sub-band?

A. A special frequency for talking to submarines
B. A frequency range limited to Extra Class licensees
C. A portion of a band where satellite operations are permitted
D. An obsolete term that has no meaning

C Because satellites can be seen anywhere in the world, it is important that all countries recognize and protect their input and output frequencies. The satellite sub-bands were developed as part of band plans worldwide so that all amateurs could use them. Sub-bands are required because satellites cannot change frequency in response to interference or noise.

T7B10 What is the satellite sub-band on 70-cm?

A. 420 to 450 MHz
B. 435 to 438 MHz
C. 440 to 450 MHz
D. 432 to 433 MHz

B 435 to 438 MHz is part of the 70-cm band plan developed by the amateur community in order to protect satellite operation from other uses. (See question T7B09.)

T7B11 What do the initials LEO tell you about an amateur satellite?

A. The satellite battery is in Low Energy Operation mode

B. The satellite is performing a Lunar Ejection Orbit maneuver

C. The satellite is in a Low Earth Orbit

D. The satellite uses Light Emitting Optics

C Low Earth Orbit is the term for satellites orbiting up to about 1600 miles above the Earth, including the International Space Station. LEO satellites have nearly circular orbits and are only visible from Earth for a few minutes. Higher orbits are referred to as Medium Earth Orbit. Farthest out are the Geosynchronous Earth Orbit satellites, whose orbital period matches exactly the rotation of the Earth so that the satellite appears to stay above one spot on Earth.

Subelement **T8**

Emergency and Public Service Communications

Your Technician exam (element 2) will consist of 35 questions taken from the Technician question pool as prepared by the Volunteer Examiner Coordinator's Question Pool Committee. A certain number of questions are taken from each of the 10 subelements. There will be 3 questions from the subelement shown in this chapter. These questions are divided into 3 groups labeled T8A through T8C.

T8A FCC declarations of an emergency, use of non-amateur equipment and frequencies, use of equipment by unlicensed persons, tactical call signs – 1 exam question

T8A01 What information is included in an FCC declaration of a temporary state of communication emergency?

A. A list of organizations authorized to use radio communications in the affected area
B. A list of amateur frequency bands to be used in the affected area
C. Any special conditions and rules to be observed during the emergency
D. An operating schedule for authorized amateur emergency stations

C [97.401(c)] - The rule states, "*When a disaster disrupts normal communication systems in a particular area, the FCC may declare a temporary state of communication emergency. The declaration will set forth special conditions and special rules to be observed by stations during the communications emergency. A request for a declaration of a temporary state of emergency should be directed to the FCC Engineer in Charge in the area concerned.*"

Here is the text of a recent communications emergency declaration: "Invoking the authority of §97.401 of the FCC's rules and regulations, Technical and Public Safety Division Chief Joseph P. Casey of the Enforcement Bureau issued a declaration that requires amateurs to refrain from using 3873 kHz during hours of darkness and 7285 kHz during hours of daylight effective immediately and until rescinded. The FCC said the communications emergency could last as long as 14 days. Both frequencies are to be protected plus or minus 3 kHz unless amateurs are taking part in the handling of emergency traffic."

T8A02 Under what conditions are amateur stations allowed to communicate with stations operating in other radio services?

A. When communicating with the space shuttle
B. When specially authorized by the FCC, or in an actual emergency
C. When communicating with stations in the Citizens Radio Service
D. When a commercial broadcast station is reporting news during a natural disaster

B [97.111(a)(2),(3),(4)] - Ordinarily amateur stations are prohibited from contacting stations in other FCC radio services or the military. (The military is not considered an FCC radio service.) There are three instances in which amateur-to-non-amateur communications are permitted.

- In support of emergency communications [97.111(a)(2)]
- While providing emergency communications as a RACES station [97.111(a)(3)]
- When specifically authorized by the FCC, such as military-to-amateur communications during Armed Forces Day celebrations [97.111(a)(4)]

T8A03 What should you do if you are in contact with another station and an emergency call is heard?

A. Tell the calling station that the frequency is in use
B. Direct the calling station to the nearest emergency net frequency
C. Disregard the call and continue with your contact
D. Stop your contact immediately and take the emergency call

D [97.101(c)] - Emergency communications always has the highest priority over other communications. Suspend your regular communications and respond to the emergency call, even if you are passing Priority or Welfare traffic or engaging in a net. Emergency calls come first!

T8A04 What are the restrictions on amateur radio communications after the FCC has declared a communications emergency?

A. The emergency declaration prohibits all communications
B. There are no restrictions if you have a special emergency certification
C. You must avoid those frequencies dedicated to supporting the emergency unless you are participating in the relief effort
D. Only military stations are allowed to use the amateur radio frequencies during an emergency

C See question T8A01. In its declaration of communications emergency the FCC can set forth any special conditions and special rules it deems necessary to support amateur emergency communications. Usually the declaration is limited to identifying frequencies to be dedicated to emergency operations.

T8A05 **What is one reason for using tactical call signs such as "command post" or "weather center" during an emergency?**

A. They help to keep the general public informed
B. They are more efficient and help coordinate public-service communications
C. They are required by the FCC
D. They increase goodwill and sound professional

B Tactical communications are performed between emergency responders or public service providers involving a few people in a small area. Stations engaged in tactical communications should use tactical call signs describing a function, location, or organization to help identify where a station is and what it is doing. Tactical calls are usually used in emergency and public service when a support or coordinating function is being provided. Tactical calls don't replace regular call signs and regular identification rules (every 10 minutes and at the end of contact) apply.

T8A06 **What is legally required to restrict a frequency to emergency-only communication?**

A. An FCC declaration of a communications emergency
B. Determination by the designated net manager for an emergency net
C. Authorization by an ARES/RACES emergency coordinator
D. A Congressional declaration of intent

A [97.401(c)] - See question T8A01.

T8A07 **Who has the exclusive use of a frequency if the FCC has not declared a communication emergency?**

A. Any net station that has traffic
B. The station first occupying the frequency
C. Individuals passing health and welfare communications
D. No station has exclusive use in this circumstance.

D Without a communications declaration in effect, no station has priority over any other station or type of activity. This is part of the flexible nature of Amateur Radio. The FCC expects amateurs to determine how best to use their spectrum, without having to develop rules for every circumstance. If you have a conflict with another station, use courtesy and common sense to resolve the problem—there's usually a way for both stations to communicate as needed.

T8A08 **What should you do if you hear someone reporting an emergency?**

A. Report the station to the FCC immediately
B. Assume the emergency is real and act accordingly
C. Ask the other station to move to a different frequency
D. Tell the station to call the police on the telephone

B Even if you think the call might be false or a prank, take it seriously. The authorities will investigate if the call turns out to be spurious. In the meantime, don't take chances and respond to the best of your ability.

T8A09 What is an appropriate way to initiate an emergency call on amateur radio?

A. Yell as loudly as you can into the microphone
B. Ask if the frequency is in use and wait for someone to give you permission to go ahead before proceeding
C. Declare a communications emergency
D. Say "Mayday, Mayday, Mayday" followed by "any station come in please" and identify your station

D Every amateur, not just sailors and pilots, should know how to make an emergency call. You might be the first on-scene at an accident or be the first to report a natural disaster. In addition, train your family members in case they need to make the call and you're not available. On CW, the proper call is "SOS SOS SOS" followed by your call sign and location.

T8A10 What are the penalties for making a false emergency call?

A. You could have your license revoked
B. You could be fined a large sum of money
C. You could be sent to prison
D. All of these answers are correct

D The authorities take false emergency calls very seriously. It wastes the time and resources of highly-trained emergency personnel. For example, it costs the Coast Guard thousands of dollars per hour to respond to a call! At the very least, you could have to repay those expenses!

T8A11 What type of communications has priority at all times in the Amateur Radio Service?

A. Repeater communications
B. Emergency communications
C. Simplex communications
D. Third-party communications

B [97.101(c)] - See question T8A03.

T8A12 When must priority be given to stations providing emergency communications?

A. Only when operating under RACES
B. Only when an emergency has been declared
C. Any time a net control station is on the air
D. At all times and on all frequencies

D [97.101(c)] - See question T8A03.

T8B Preparation for emergency operations, RACES/ARES, safety of life and property, using ham radio at civic events, compensation prohibited – 1 exam question

T8B01 What can you do to be prepared for an emergency situation where your assistance might be needed?

A. Check at least twice a year to make sure you have all of your emergency response equipment and know where it is

B. Make sure you have a way to run your equipment if there is a power failure in your area

C. Participate in drills that test your ability to set up and operate in the field

D. All of these answers are correct

D The very first item in the Basis and Purpose for the Amateur service [97.1] is providing emergency communications. To be effective, you must also be prepared. Preparedness means having the resources, both radio equipment and power to run it, and the know-how to communicate properly. The best way to be prepared is to treat drills and exercises as a real emergency. Assemble a "go-kit" of radio and personal gear, then use it whenever you're given the opportunity.

T8B02 When may you use your amateur station to transmit an "SOS" or "MAYDAY" signal?

A. Only when you are transmitting from a ship at sea

B. Only at 15 and 30 minutes after the hour

C. When there is immediate threat to human life or property

D. When the National Weather Service has announced a weather warning

C [97.403] - Emergencies have a way of happening in unexpected circumstances, so the FCC does not place any conditions on distress signals. Rule [97.403] is clear in stating that you may use whatever means necessary, including transmitting SOS and MAYDAY signals, regardless of the rules for normal communications.

T8B03 What is the primary function of RACES in relation to emergency activities?

A. RACES organizations are restricted to serving local, state, and federal government emergency management agencies

B. RACES supports agencies like the Red Cross, Salvation Army, and National Weather Service

C. RACES supports the National Traffic System

D. RACES is a part of the National Emergency Warning System

A RACES stands for Radio Amateur Civil Emergency Service. RACES is sponsored by local and state civil defense organizations and supported by the Federal Emergency management Administration (FEMA), operating as a separate service within the Part 97 rules. RACES is intended to provide government-to-government communications for civil defense purposed only, during periods of local, regional, or national civil emergencies. RACES operation is strictly limited to official civil defense activities in an emergency communications situation.

T8B04 **What is the primary function of ARES in relation to emergency activities?**

A. ARES organizations are restricted to serving local, state, and federal government emergency management agencies
B. ARES supports agencies like the Red Cross, Salvation Army, and National Weather Service
C. ARES groups work only with local school districts
D. ARES supports local National Guard units

B ARES stands for the Amateur Radio Emergency Service and consists of amateur volunteers that provide communications for public service of all kinds, not just emergency communications. ARES volunteers work with both government and non-government organizations. The ARES is sponsored by the ARRL and is organized as part of the ARRL's Field Services department.

T8B05 **What organization must you register with before you can participate in RACES activities?**

A. A local amateur radio club
B. A local racing organization
C. The responsible civil defense organization
D. The Federal Communications Commission

C [97.407(a)] - No station may transmit in RACES unless it is an FCC-licensed primary, club, or military recreation station and it is certified by a civil defense organization and registered with that organization. In addition, RACES participation is limited to communications with other RACES stations, government stations, and stations in other services as authorized by the FCC.

T8B06 **What is necessary before you can join an ARES group?**

A. You are required to join the ARRL
B. You must have an amateur radio license
C. You must have an amateur radio license and have Red Cross CPR training
D. You must register with a civil defense organization

B Any licensed amateur can be a member of ARES. Although it is organized by the ARRL, ARES membership is open to all. Aside from a license, the only requirement is a genuine desire to provide service.

T8B07 **What could be used as an alternate source of power to operate radio equipment during emergencies?**

A. The battery in a car or truck
B. A bicycle generator
C. A portable solar panel
D. All of these answers are correct

D Since many amateur radios operate from 12 V dc, the electrical system of a car or truck is an excellent backup power source. It's a good idea to make provisions for powering your radio from your vehicle before the emergency, so it's easy for you make the transition when needed. Bicycle generators, solar panels, and even windmills are alternative sources of power!

T8B08 When can you use non-amateur frequencies or equipment to call for help in a situation involving immediate danger to life or property?

A. Never; your license only allows you to use the frequencies authorized to your class of license
B. In a genuine emergency you may use any means at your disposal to call for help on any frequency
C. When you have permission from the owner of the set
D. When you have permission from a police officer on the scene

B [97.403, 97.405(a),(b)] - Stations in distress should use any means at their disposal to attract attention and obtain assistance. This includes operating on frequencies and equipment for which they are not licensed.

T8B09 Why should casual conversation between stations during a public service event be avoided?

A. Such chatter is often interesting to bystanders
B. Other listeners might overhear personal information
C. Idle chatter may interfere with important traffic
D. You might have to change batteries more often

C Radio discipline is an important part of effective emergency and public service communications. Even though nothing is happening where you are, another station may need assistance or have important traffic. Don't tie up the frequency with casual conversation--keep your transmissions brief and focused on the business at hand.

T8B10 What should you do if a reporter asks to use your amateur radio transceiver to make a news report?

A. Allow the use but give your call sign every 10 minutes
B. Advise them that the FCC prohibits such use
C. Tell them it is OK as long as you do not receive compensation
D. Tell the reporter that you must approve the material beforehand

B The FCC rules are quite clear in that Amateur Radio is not to be used for collecting or relaying messages in support of broadcast news organizations. The desire for information in emergency and disaster situations is intense. To keep the Amateur Radio service from being co-opted as a way for news organizations to gather information, the FCC Rules are clear—no news traffic!

T8B11 When can you use a modified amateur radio transceiver to transmit on the local fire department frequency?

A. When you are helping the Fire Department raise money
B. Only when the Fire Department is short of regular equipment
C. In a genuine emergency you may use any means at your disposal to call for help on any frequency
D. When the local Fire Chief has given written permission

C [97.403, 97.405(a),(b)] Most radio services are only permitted to use *type-accepted* radio equipment that is certified to meet the standards for that service. Amateur Radio equipment is not certified for use on any other service and should not be used *except* in the case of emergencies with a threat to life and property.

T8C Net operations, responsibilities of the net control station, message handling, interfacing with public safety officials - 1 exam question

T8C01 Which type of traffic has the highest priority?

A. Emergency traffic
B. Priority traffic
C. Health and welfare traffic
D. Routine traffic

A Just as emergency communications has the highest priority, so do emergency messages have the highest priority. If a station has emergency traffic, all other messages and communications should be put on hold until the emergency traffic has been successfully relayed.

T8C02 What type of messages should not be transmitted over amateur radio frequencies during emergencies?

A. Requests for supplies
B. Personal information concerning victims
C. A schedule of relief operators
D. Estimates of how much longer the emergency will last

B It is important that sensitive personal information about victims not be disclosed. If you are asked to transmit that information, decline or use a secure method of transmission not easily overhead by the public.

T8C03 **What should you do to minimize disruptions to an emergency traffic net once you have checked in?**

A. Whenever the net frequency is quiet, announce your call sign and location

B. Move 5 kHz away from the net's frequency and use high power to ask other hams to keep clear of the net frequency

C. Do not transmit on the net frequency until asked to do so by the net control station

D. Wait until the net frequency is quiet, then ask for any emergency traffic for your area

C The desire to help is strong, but remember that unnecessary transmissions just slow things down. The Net Control Station (NCS) will record your call sign and location so that if you're needed, you can be called. There's no need to remind the NCS that you're listening. Once you've checked in, stay on frequency until you check out.

T8C04 **What is one thing that must be included when passing emergency messages?**

A. The call signs of all the stations passing the message

B. The name of the person originating the message

C. A status report

D. The message title

B It is important to be able to trace messages back to their source. While having the call signs of relaying stations is helpful, the most important piece of information for tracing messages is the name of the person originating the message. This is not the name or call sign of the operator who first sent the message, it is the original source of the *information*. This also helps the message recipient respond to the message.

T8C05 **What is one way to reduce the chances of casual listeners overhearing sensitive emergency traffic?**

A. Pass messages using a non-voice mode such as packet radio or Morse code

B. Speak as rapidly as possible to reduce your on-air time

C. Spell out every word using phonetics

D. Restrict transmission of messages to the hours between midnight and 4:00 AM

A If amateur radio is the only means of communication available for sensitive information such as names or addresses, pass the information using a mode of communications less likely to be received by the public, such as packet radio or Morse code.

T8C06 What is of primary importance for a net control station?

A. A dual-band transceiver
B. A network card
C. A strong and clear signal
D. The ability to speak several languages

C The Net Control Station (NCS) needs to be clearly understood by all net stations. Everyone needs to hear the NCS' instructions and announcements. Having a strong signal also prevents confusion that occurs when stations are not quite sure of when the NCS is transmitting.

T8C07 What should the net control station do if someone breaks in with emergency traffic?

A. Ask them to wait until the roll has been called
B. Stop all net activity until the emergency has been handled
C. Ask the station to call the local police and then resume normal net activities
D. Ask them to move off your net frequency immediately

B See question T8C01.

T8C08 What should you do if a large scale emergency has just occurred and no net control station is available?

A. Wait until the assigned net control station comes on the air and pass your traffic when called
B. Transmit a call for help and hope someone will hear you
C. Open the emergency net immediately and ask for check-ins
D. Listen to the local NOAA weather broadcast to find out how long the emergency will last

C If you are the first one on the air after a disaster or emergency, you can establish a net on the frequency and take check-ins. That way, if you are uncomfortable acting as a full Net Control Station, when a suitable station is ready, you can provide them with a list of stations, their locations, and any messages or traffic. This prevents the NCS from having to check everyone in again.

T8C09 What is the preamble of a message?

A. The first paragraph of the message text
B. The message number
C. The priority handling indicator for the message
D. The information needed to track the message as it passes through the amateur radio traffic handling system

D Each message needs some information *about* the message and its content. The preamble gives the message a unique identity. The preamble consists of the following information:

- A message number
- The type of message
- Handling instructions for the message, if any
- Information about the length of the message to help detect errors (the **check**)
- The station that first accepted the information and started the message on its way
- When and where the message was created.

T8C10 What is meant by the term "check" in reference to a message?

A. The check is a count of the number of words in the message
B. The check is the value of a money order attached to the message
C. The check is a list of stations that have relayed the message
D. The check is a box on the message form that tells you the message was received

A Knowing the length of the message gives the receiving station an opportunity to make sure it was received in its entirety. The check is a count of all the words and numbers *between* the address of the message and the signature, if any.

T8C11 What is the recommended guideline for the maximum number of words to be included in the text of an emergency message?

A. 10 words
B. 25 words
C. 50 words
D. 75 words

B A lot of information can be packed into a 25-word message! By limiting messages to 25 words, the information becomes quite focused and the messages are easier to transmit reliably. If longer messages need to be sent, such as lists, it would be better to use a more suitable mode, such as packet radio or Winlink.

Subelement **T9**

Antenna Systems and Propagations

Your Technician exam (element 2) will consist of 35 questions taken from the Technician question pool as prepared by the Volunteer Examiner Coordinator's Question Pool Committee. A certain number of questions are taken from each of the 10 subelements. There will be 3 questions from the subelement shown in this chapter. These questions are divided into 3 groups labeled T9A through T9C.

T9A Antenna types – vertical, horizontal, concept of gain, common portable and mobile antennas, losses with short antennas, relationships between antenna length and frequency, dummy loads 1 exam question

T9A01 What is a beam antenna?

A. An antenna built from metal I-beams
B. An antenna that transmits and receives equally well in all directions
C. An antenna that concentrates signals in one direction
D. An antenna that reverses the phase of received signals

C The term "beam" is used in the same sense as a flashlight beam. In fact, the reflector of a flashlight acts to concentrate the light bulb's "signal" in one direction, just as a radio-frequency beam antenna does. The beam doesn't create more power; it just focuses it.

T9A02 What is an antenna that consists of a single element mounted perpendicular to the Earth's surface?

A. A conical monopole
B. A horizontal antenna
C. A vertical antenna
D. A traveling wave antenna

C The vertical antenna has a pair of reasons for the name. First, the antenna is, in fact, vertically oriented! Second, the vertical orientation of the antenna element means that the electric field component of its radiated waves is also vertical.

T9A03 What type of antenna is a simple dipole mounted so the elements are parallel to the Earth's surface?

A. A ground wave antenna
B. A horizontal antenna
C. A rhombic antenna
D. A vertical antenna

B Like the vertical, a dipole is called a horizontal antenna for two reasons. The antenna element is in a horizontal position and the radiated radio wave's electric field is also oriented horizontally.

T9A04 What is a disadvantage of the "rubber duck" antenna supplied with most hand held radio transceivers?

A. It does not transmit or receive as effectively as a full sized antenna
B. It is much more expensive than a standard antenna
C. If the rubber end cap is lost it will unravel very quickly
D. It transmits a circular polarized signal

A When you buy a new VHF or UHF hand-held transceiver, it will usually have a flexible rubber-coated antenna commonly called a "rubber duck." This antenna is inexpensive, small, lightweight, and difficult to break. On the other hand, its performance is not nearly as good as a full-sized antenna, such as a mobile antenna or a telescoping whip.

T9A05 How does the physical size of half-wave dipole antenna change with operating frequency?

A. It becomes longer as the frequency increases
B. It must be made larger because it has to handle more power
C. It becomes shorter as the frequency increases
D. It becomes shorter as the frequency decreases

C Remember that wavelength and frequency change in opposite directions; as one increases, the other decreases and vice versa. If frequency doubles, wavelength is halved. This is because of the constant speed of the radio waves.

T9A06 What is the advantage of ⅝ wavelength over ¼ wavelength vertical antennas?

A. They are easier to match to the feed line than other types
B. Their radiation pattern concentrates energy at lower angles
C. They pick up less noise
D. Their radiation pattern concentrates energy at higher angles

B The longer element of the ⅝-wavelength antenna allows it to concentrate more of the radiated energy at right angles to its length. Because the antenna is oriented vertically, this means more the radiated energy is focused horizontally, where other stations are, improving the received signal.

T9A07 What is the primary purpose of a dummy load?

A. It does not radiate interfering signals when making tests
B. It will prevent over-modulation of your transmitter
C. It keeps you from making mistakes while on the air
D. It is used for close in work to prevent overloads

A Because the dummy load absorbs all of the transmitter output and radiates it as heat instead of radio waves, there is no signal to interfere with other hams. You should always use a dummy load when testing a transmitter unless you absolutely have to make an on-the-air test.

T9A08 What type of antennas are the quad, Yagi, and dish?

A. Antennas invented after 1985
B. Loop antennas
C. Directional or beam antennas
D. Antennas that are not permitted for amateur radio stations

C All of these antennas focus the radiated energy towards one direction. The quad and Yagi work by having groups of elements work together to reinforce radiated energy in the desired direction. The dish antenna works like a flashlight, reflecting energy so that a great deal of it is focused in one direction.

T9A09 What is one type of antenna that offers good efficiency when operating mobile and can be easily installed or removed?

A. A microwave antenna
B. A quad antenna
C. A traveling wave antenna
D. A magnet mount vertical antenna

D "Magnet mount" or "mag mount" refers to the method of attaching the antenna to the vehicle. The antenna in mounted on a large, circular magnet that attaches to the steel body of the vehicle. The vehicle forms the ground plane for the antenna The magnet is strong enough to keep the antenna on the car at highway speeds, but it easy to remove for parking and in the garage.

T9A10 What is a good reason not to use a "rubber duck" antenna inside your car?

A. Signals can be 10 to 20 times weaker than when you are outside of the vehicle
B. RF energy trapped inside the vehicle can distort your signal
C. You might cause a fire in the vehicle upholstery
D. The SWR might increase

A The vehicle's metal roof and doors act like shields, trapping the radio waves inside. Some of the signal gets out through the windows (unless they're tinted by a thin metal coating), but it's much weaker than if radiated by an external antenna.

T9A11 What is the approximate length, in inches, of a quarter-wavelength vertical antenna for 146 MHz?

A. 112 inches
B. 50 inches
C. 19 inches
D. 12 inches

C Start with the formula for the length of a half-wavelength dipole:

Length (in feet) = 468 / frequency (in MHz)

In this case, length = 468 / 146 = 3.21 feet. Convert to inches by multiplying by 12 = 38.47. Because this is a quarter-wavelength antenna, divide the result by two to get 19.23 or about 19¼".

T9A12 What is the approximate length, in inches, of a 6-meter ½ wavelength wire dipole antenna?

A. 6 inches
B. 50 inches
C. 112 inches
D. 236 inches

C Since you already know the half-wavelength (3 meters), convert meters to feet by multiplying by 3.1 = 9.3 feet. Convert to inches by multiplying by 12 = 111.6".

T9B Propagation, fading, multipath distortion, reflections, radio horizon, terrain blocking, wavelength vs penetration, antenna orientation – 1 exam question

T9B01 Why are VHF/UHF signals not normally heard over long distances?

A. They are too weak to go very far
B. FCC regulations prohibit them from going more than 50 miles
C. VHF and UHF signals are usually not reflected by the ionosphere
D. They collide with trees and shrubbery and fade out

C Because the frequency of VHF/UHF signals is too high for the ionosphere to reflect, most VHF/UHF signals are limited to *line-of-sight* propagation, meaning that they travel in straight lines. Receiving antennas past where the Earth's curvature blocks the transmitting antenna are out of range for these signals.

T9B02 What might be happening when we hear a VHF signal from long distances?

A. Signals are being reflected from outer space
B. Someone is playing a recording to us
C. Signals are being reflected by lightning storms in our area
D. A possible cause is sporadic E reflection from a layer in the ionosphere

D Most common on the 6-meter, 2-meter, and 1¼-meter bands, **sporadic-E** propagation occurs from reflections off highly ionized patches in the ionosphere's E-layer. The patches form at irregular intervals and places, earning them the name "sporadic." Sporadic-E propagation can relay signals as far as 1500 miles.

T9B03 What is the most likely cause of sudden bursts of tones or fragments of different conversations that interfere with VHF or UHF signals?

A. The batteries in your transceiver are failing
B. Strong signals are overloading the receiver and causing undesired signals to be heard
C. The receiver is picking up low orbit satellites
D. A nearby broadcast station is having transmitter problems

B Overload often occurs when a receiver is operated in the vicinity of strong signals, such as on hills near transmitter sites, buildings with transmitter is located on their roofs, and similar locations. A repeater receiver, if overloaded, will also transmit the spurious signals through the repeater transmitter. The solution is usually better filtering at the receiver to reject the strong, unwanted signals.

T9B04 **What is the radio horizon?**

A. The point where radio signals between two points are blocked by the curvature of the Earth
B. The distance from the ground to a horizontally mounted antenna
C. The farthest point you can see when standing at the base of your antenna tower
D. The shortest distance between two points on the Earth's surface

A Since most propagation at VHF and UHF frequencies is line-of-sight, the limit of the range of these signals is determined by the radio horizon. (There is some slight bending of radio waves along the Earth's surface, so the radio horizon is slightly more distant than the visual horizon.) Increasing the height of either the transmitting or receiving antennas also increases the radio horizon's distance.

T9B05 **What should you do if a station reports that your signals were strong just a moment ago, but now they are weak or distorted?**

A. Change the batteries in your radio to a different type
B. Speak more slowly so he can understand your better
C. Ask the other operator to adjust his squelch control
D. Try moving a few feet, random reflections may be causing multipath distortion.

D In a location where reflections can occur, such as near vehicles or buildings, it is possible for a direct signal and a reflected signal to interfere with each other at the receiver and partially cancel. When this occurs, all that is necessary is to move approximately one-half wavelength in any direction; about 3 feet on the 2-meter band. It is likely that you'll find a "hot spot" within that range, or at least a location where the cancellation is greatly reduced.

T9B06 **Why do UHF signals often work better inside of buildings than VHF signals?**

A. VHF signals lose power faster over distance
B. The shorter wavelength of UHF signals allows them to more easily penetrate urban areas and buildings
C. This is incorrect; VHF works better than UHF inside buildings
D. UHF antennas are more efficient than VHF antennas

B Because a UHF signal's wavelength is shorter, obstructions and holes look larger in terms of wavelength. A larger hole will allow more signal through, so UHF signals get in and out of buildings and vehicles more easily. Larger obstructions also *diffract* or bend signals better, so UHF signals also have less *shadowing* effects in urban areas.

T9B07 **What is a good thing to remember when using your hand-held VHF or UHF radio to reach a distant repeater?**

A. Speak as loudly as possible to help your signal go farther
B. Keep your transmissions short to conserve battery power
C. Keep the antenna as close to vertical as you can
D. Turn off the CTCSS tone

C On the VHF and UHF bands, it is important to keep the transmitting and receiving antennas aligned so that they have matching polarizations. If the radio wave from a transmitter arrives at the receiver with a different polarization, the receiving antenna does not respond as well to the incoming radio wave. Since most repeaters have vertically-polarized antennas, it is important that your antenna be vertically polarized, too.

T9B08 **What can happen if the antennas at opposite ends of a VHF or UHF line of sight radio link are not using the same polarization?**

A. The modulation sidebands might become inverted
B. Signals could be as much as 100 times weaker
C. Signals have an echo effect on voices
D. Nothing significant will happen

B See question T9B07.

T9B09 **What might be a way to reach a distant repeater if buildings or obstructions are blocking the direct line of sight path?**

A. Change from vertical to horizontal polarization
B. Try using a directional antenna to find a path that reflects signals to the repeater
C. Ask the repeater owners to repair their receiver
D. Transmit on the repeater output frequency

B Remember that building and hills can act as radio reflectors. If you can "aim" your signal at one of these large reflectors, it is often possible to evade an obstruction directly in your line of sight to the desired station.

T9B10 **What term is commonly used to describe the rapid fluttering sound sometimes heard from mobile stations that are moving while transmitting?**

A. Flip-flopping
B. Picket fencing
C. Frequency shifting
D. Pulsing

B See question T9B06. Imagine yourself driving through an area with lots of reflection. You will travel in and out of "hot spots" and "dead spots" quite rapidly. The effect on your signal is an equally rapid increase and decrease in signal strength. Also called **mobile flutter**, the distinctive result sounds like a stick being dragged along a picket fence.

T9B11 **Why do VHF and UHF Radio signals usually travel about a third farther than the visual line of sight distance between 2 stations?**

A. Radio signals move somewhat faster than the speed of light and travel farther in the same amount of time
B. Radio waves are not blocked by dust particles
C. The Earth seems less curved to radio waves than to light
D. Radio waves are blocked by dust particles

C Because of this slight bending of the radio waves as they travel through the atmosphere above the Earth, the radio horizon is about a third more distant than that visual horizon. See also question T9B04.

T9C Feedlines types, losses vs frequency, SWR concepts, measuring SWR, matching and power transfer, weather protection, feedline failure modes 1 exam question

T9C01 What, in general terms, is standing wave ratio (SWR)?

A. A measure of how well a load is matched to a transmitter
B. The ratio of high to low impedance in a feed line
C. The transmitter efficiency ratio
D. An indication of the quality of your station ground connection

B Standing wave ratio, or SWR, is caused by a mismatch of the characteristic impedance of the feed line and that of a load (such as an antenna) attached to the feed line. Some of the energy delivered by the feed line is reflected at the load and interferes with the incoming energy, causing standing waves The magnitude of the SWR is the ratio of the higher of the two impedances to the lower.

T9C02 What reading on a SWR meter indicates a perfect impedance match between the antenna and the feed line?

A. 2 to 1
B. 1 to 3
C. 1 to 1
D. 10 to 1

C If the feed line impedance and load impedance are the same, their ratio is 1:1. It can't get any better than that!

T9C03 What might be indicated by erratic changes in SWR readings?

A. The transmitter is being modulated
B. A loose connection in your antenna or feed line
C. The transmitter is being over modulated
D. Interference from other stations is distorting your signal

B Remember that SWR is caused by a mismatch of feed line impedance and load impedance. If there is a loose connection where the load is attached (i.e. at the antenna feed point), then the erratic connection acts like an erratically changing load impedance. The result is an erratic change in SWR.

T9C04 What is the SWR value where the protection circuits in most solid-state transmitters begin to reduce transmitter power?

A. 2 to 1
B. 1 to 2
C. 6 to 1
D. 10 to 1

A The standing waves produced by the reflected energy result in increased voltages at points along the feed line. This higher voltage, if present at the transmitter output, can damage the output circuits of the transmitter. Protective circuitry senses SWR and when it becomes excessive, reduces the amount of power being supplied to the line, reducing the strength of the reflected power and the increased voltage it causes. An SWR of 2 to 1 (2:1) is the usual point at which protection becomes necessary.

T9C05 What happens to the power lost in a feed line?

A. It increases the SWR
B. It comes back into your transmitter and could cause damage
C. It is converted into heat by losses in the line
D. It can cause distortion of your signal

C As energy travels through a feed line, some of it is absorbed by the insulation in the line and some is lost in the resistance of the conductors themselves. These *losses* are in the form of heat, just as if the energy was dissipated by a resistor.

T9C06 What instrument other than a SWR meter could you use to determine if your feed line and antenna are properly matched?

A. Voltmeter
B. Ohmmeter
C. Iambic pentameter
D. Directional wattmeter

D A directional wattmeter can tell you how much energy is flowing in each direction along a feed line. From the relative amounts of power flowing in each direction, you can calculate SWR. By minimizing reflected power, you can adjust your antenna for the best impedance match.

T9C07 What is the most common reason for failure of coaxial cables?

A. Moisture contamination
B. Gamma rays
C. End of service life
D. Overloading

A Capillary action of the strands making up the braided outer shield can draw water into a flexible coaxial cable. Once in the cable, the water causes both corrosion and heat losses in the shield. Water can get into all coaxial cables either through cracks or tears in the plastic outer jacket or through an improperly sealed connection.

T9C08 **Why is it important to have a low SWR in an antenna system that uses coaxial cable feed line?**

A. To reduce television interference
B. To allow the efficient transfer of power and reduce losses
C. To prolong antenna life
D. To keep your signal from changing polarization

B Power reflected from a mismatched antenna bounces back and forth along the feed line, a bit more being transferred to the antenna each time, but each trip exacts some loss. At high SWR, a significant fraction of the transmitter output power can be lost.

T9C09 **What can happen to older coaxial cables that are exposed to weather and sunlight for several years?**

A. Nothing, weather and sunlight do not affect coaxial cable
B. The cable can shrink and break
C. Losses can increase dramatically
D. It will short-circuit

C Exposure to weather and sunlight can cause cracks in the outer plastic jacket or shrinkage and peeling of sealing materials at connectors. Both can let water get into the cable--see question T9C07.

T9C10 **Why is the outer sheath of most coaxial cables black in color?**

A. It is the cheapest color to use
B. To see nicks and cracks in the cable
C. Black cables have less loss
D. Black provides protection against ultraviolet damage

D The pigment in black plastic often acts to block or absorb ultraviolet light, protecting the rest of the jacket.

T9C11 **What is the impedance of the most commonly used coaxial cable in typical amateur radio installations?**

A. 8 Ohms
B. 50 Ohms
C. 600 Ohms
D. 12 Ohms

B Radio manufacturers standardized on 50-ohms for most coaxial cables by 1950. Fifty ohms was chosen because cables with that impedance have a combination of good power handling and voltage rating capability.

T9C12 Why is coaxial cable used more often than any other feed line for amateur radio antenna systems?

A. It is easy to use and requires few special installation considerations
B. It has less loss than any other type of feed line
C. It can handle more power than any other type of feed line
D. It is less expensive than any other types of line

A Coaxial cable, or *coax*, is easy to work with because it is a single cable, like rope. All of the energy being conducted by the feed line is completely contained within the cable, so the cable can be run along side or even within metallic trays, conduits, and towers. It can be coiled up and placed next to other cables without effect. While it may have more loss than open-wire line and weigh a bit more per foot, coaxial cable's ease of use make it a practical choice for most installations.

Subelement T0

Safety

Your Technician exam (element 2) will consist of 35 questions taken from the Technician question pool as prepared by the Volunteer Examiner Coordinator's Question Pool Committee. A certain number of questions are taken from each of the 10 subelements. There will be 3 questions from the subelement shown in this chapter. These questions are divided into 3 groups labeled T0A through T0B.

T0A AC power circuits, hazardous voltages, fuses and circuit breakers, grounding, lightning protection, battery safety, electrical code compliance – 1 exam question

T0A01 **What is a commonly accepted value for the lowest voltage that can cause a dangerous electric shock?**

A. 12 volts
B. 30 volts
C. 120 volts
D. 300 volts

B Low-voltage power supplies may seem safe, but even battery-powered equipment should be treated with care. The minimum voltage that considered a shock hazard to humans is 30 volts. You should respect even low voltages, taking appropriate steps to avoid contact. Lower-voltage systems such as storage batteries are quite capable of causing a fire hazard, so even without being a shock hazard, there are plenty of reasons to treat them with care.

Table T0-1

Effects of Electric Current Through the Body of an Average Person

Current (1 Second Contact)	*Effect*
1 mA	Barely noticeable
5 mA	Maximum harmless current
10-20 mA	Lower limit for sustained muscle contractions
30-50 mA	Pain
50 mA	Pain and possible loss of consciousness
100-300 mA	Heart rhythm disrupted. Death possible
6 A	Heart rhythm severely disrupted. Burns. Death.

T0A02 What is the lowest amount of electrical current flowing through the human body that is likely to cause death?

A 10 microamperes
B. 100 milliamperes
C. 10 amperes
D. 100 amperes

B You should never underestimate the potential hazard when working with electricity. As little as 100 milliamps (mA), or 1/10th amp (A), can be fatal! As the saying goes, "It's the volts that jolts, but it's the mills that kills." Table T0-1 shows the effects on the human body at various levels of current.

T0A03 What is connected to the green wire in a three-wire electrical plug?

A. Neutral
B. Hot
C. Ground
D. The white wire

C State and national electrical-safety codes require the three-wire power cords on many 120-V tools and appliances. Power supplies and station equipment use similar connections. Two of the conductors (the "hot" and the "neutral" wires) power the device. The third conductor (the safety ground wire) connects to the metal frame of the device. The "hot" wire is usually black or red. The "neutral" wire is white. The frame/ground wire is green or sometimes bare.

T0A04 What is the purpose of a fuse in an electrical circuit?

A. To make sure enough power reaches the circuit
B. To interrupt power in case of overload
C. To prevent television interference
D. To prevent shocks

B A fuse consists of a thin strip of metal that melts at relatively low temperatures. Current flow causes the metal to heat up and when too much current flows, the metal melts, breaking the circuit and interrupting power.

T0A05 **What might happen if you install a 20-ampere fuse in your transceiver in the place of a 5-ampere fuse?**

A. The larger fuse would better protect your transceiver from using too much current
B. The transceiver will run cooler
C. Excessive current could cause a fire
D. The transceiver would not be able to produce as much RF output

C Never replace a fuse or circuit breaker with one rated for a larger current, since that allows more current to flow in response to a fault in the equipment. The higher current could overheat wires and cause a fire. Determine what problem caused the fuse to blow and repair it before higher current makes the damage worse or destroys the equipment entirely.

T0A06 **What is a good way to guard against electrical shock at your station?**

A. Use 3-wire cords and plugs for all AC powered equipment
B. Connect all AC powered station equipment to a common ground
C. Use a ground-fault interrupter at each electrical outlet
D. All of these answers are correct

D The best way to protect against electrical shock is to make sure that each part of your station as at the same potential or voltage. That potential should, of course, be ground potential. You'll want to make sure that you have a common ground, meaning that all grounds are connected together. Three-wire cords and plugs make sure the equipment enclosures are connected to the common ground. A ground-fault interrupter detects current imbalance between the hot and neutral wires, indicating a short-circuit to the common ground, and turns off power to the circuit.

T0A07 **What is the most important thing to consider when installing an emergency disconnect switch at your station?**

A. It must always be as near to the operator as possible
B. It must always be as far away from the operator as possible
C. Everyone should know where it is and how to use it
D. It should be installed in a metal box to prevent tampering

C Anyone discovering a person being shocked or burned by high voltage should immediately turn off the power, call for help and give cardiopulmonary resuscitation (CPR) if needed. To turn off the power safely, the master power disconnect switch should be clearly labeled and located where everyone can use it. Train all family members how to turn off power using the switch. These measures could save a life!

T0A08 What precautions should be taken when a lightning storm is expected?

A. Disconnect the antenna cables from your station and move them away from your radio equipment
B. Unplug all power cords from AC outlets
C. Stop using your radio equipment and move to another room until the storm passes
D. All of these answers are correct

D Disconnecting and grounding all antennas, feed lines, and rotator cables is an effective means of lightning protection. Grounding prevents an antenna from building up an electrical charge from a nearby storm that can be damaging to radios, particularly sensitive receivers. You should also unplug your equipment (both power and telephone cords) since a lightning strike on a power line can conduct the strike into your home. Power and telephone lines can act as antennas, picking up significant amounts of energy even if not struck directly. To be protected, equipment must be completely unplugged, since lightning can easily jump across a circuit breaker. To protect yourself, stay away from your radio equipment until the storm passes.

T0A09 What is one way to recharge a 12-volt battery if the commercial power is out?

A. You cannot recharge a battery unless the power is back on
B. Add water to the battery
C. Connect the battery to a car's battery and run the engine
D. Take your battery to the utility company for a recharge

C Your vehicle is an excellent source of backup power. It can easily recharge a storage battery. Smaller batteries should be carefully monitored for excessive temperature when charging from a vehicle since there is no circuitry to limit charging current. Follow standard vehicle safety practices when using your car to recharge batteries.

T0A10 What kind of hazard is presented by a conventional 12-volt storage battery?

A. It contains dangerous acid that can spill and cause injury
B. Short circuits can damage wiring and possibly cause a fire
C. Explosive gas can collect if not properly vented
D. All of these answers are correct

D Storage batteries pack a lot of energy into a small volume, but with that energy comes the need to treat them carefully. The liquid acid in the battery is extremely corrosive and will eat holes in anything organic (including your clothes and skin!) During charging, hydrogen gas is given off by the battery and can be explosive if there is no ventilation to disperse it. Most of all, a short-circuit in a storage battery circuit can draw hundreds of amps from the battery, melting wires and terminals and igniting insulation in a hurry! Take special care to keep tools from accidentally shorting across the battery terminals!

T0A11 **What can happen if a storage battery is charged or discharged too quickly?**

A. The battery could overheat and give off dangerous gas or explode
B. The terminal voltage will oscillate rapidly
C. The warranty will be voided
D. The voltage will be reversed

A Trying to get energy into or out of any battery too quickly can cause it to overheat. Excessive charging can cause hydrogen gas to build up faster than the battery can vent it to the outside air. Overheating is particularly dangerous with storage batteries, since they hold so much energy. Use a battery charger that is designed to work with storage batteries and keep the load on the battery to within the manufacturer's specifications.

T0A12 **What is the most important reason to have a lightning protection system for your amateur radio station?**

A. Lower insurance rates
B. Improved reception
C. Fire prevention
D. Noise reduction

C Lightning strikes carry thousands of amperes of current. Even an indirect strike can severely overload your station wiring and connectors. By protecting your station against lightning strikes, you also protect yourself against the possibility of fires started by a strike.

T0A13 **What kind of hazard might exist in a power supply when it is turned off and disconnected?**

A. Static electricity could damage the grounding system
B. Circulating currents inside the transformer might cause damage
C. The fuse might blow if you remove the cover
D. You might receive an electric shock from stored charge in large capacitors

D A capacitor is designed to store charge, so it should not be a surprise that a capacitor can remain charged even after power is removed. It is good practice to place *bleeder resistors* across large capacitors to slowly allow the charge to dissipate. Never assume that capacitors, particularly those in high-voltage circuits, are discharged. Measure them with a voltmeter first.

T0B Antenna installation, tower safety, overhead power lines – 1 exam question

T0B01 Why should you wear a hard hat and safety glasses if you are on the ground helping someone work on an antenna tower?

A. It is required by FCC rules
B. To keep RF energy away from your head during antenna testing
C. To protect your head and eyes in case something accidentally falls from the tower
D. It is required by the electrical code

C A piece of hardware or a tool will be traveling 40 mph by the time it falls 60 feet! Ouch! Whenever the crew is working on the tower, wear your protective gear—even if you are near the base of the tower. If a falling object hits the tower or a guy wire, it can bounce a long way. Try to stay out from under the crew at work at the top. And while you're at it, remember the sun block!

T0B02 What is a good precaution to observe before climbing an antenna tower?

A. Turn on all radio transmitters
B. Remove all tower grounding connections
C. Put on your safety belt and safety glasses
D. Inform the FAA and the FCC that you are working on a tower

C Make sure your belt and glasses and hardhat are in good condition—then use them! Climb slowly, with a lanyard around the tower. It's not a race, so take your time. Once at the top, work slowly, thinking out each move before you make it. Have a backup plan and never work on a tower without a ground crew or someone to keep an eye on you.

T0B03 What should you do before you climb a tower?

A. Arrange for a helper or observer
B. Inspect the tower for damage or loose hardware
C. Make sure there are no electrical storms nearby
D. All of these answers are correct

D Before you climb a tower, be sure someone knows you will be up there. Make sure you can get help quickly if it is needed. Before you take the first step up the tower, inspect it for loose hardware and cracks. Use binoculars for the higher hardware. Inspect the guy wires and anchors for loose or fraying cable, loose or cracked turnbuckles, or slipping or loose clamps. All of these items are critical safety issues. Make every item a part of your regular routine before a climb.

T0B04 What is an important consideration when putting up an antenna?

A. Carefully tune it for a low SWR
B. Make sure people cannot accidentally come into contact with it
C. Make sure you discard all packing material in a safe place
D. Make sure birds can see it so they don't fly into it

B Antennas can present both electrical and mechanical hazards. Even without using an amplifier, an antenna could have enough RF voltage on it to deliver a painful RF burn. Tripping on or being snagged by an antenna or feed line are also unwelcome surprises. Install your antenna and feed line to stay well clear of people and animals.

T0B05 What must be considered when erecting an antenna near an airport?

A. The maximum allowed height with regard to nearby airports
B. The possibility of interference to aircraft radios
C. The radiation angle of the signals it produces
D. The polarization of signal to be radiated

A [97.15(A)] - Owners of towers more than 200 feet high or located near or at a public use airport must notify the Federal Aviation Administration and register with the FCC. High towers are then marked on aviation charts. Near airports, a tall tower may present a flight safety hazard, so it's important to site the tower carefully.

T0B06 What is the most important safety precaution to observe when putting up an antenna tower?

A. Install steps on the tower for safe climbing
B. Insulate the base of the tower to avoid lightning strikes
C. Ground the base of the tower to prevent lightning strikes
D. Look for and stay clear of any overhead electrical wires

D If power lines ever come into contact with your antenna, you could be electrocuted. The only safe place to install an antenna tower is in a location that is well clear of any power lines. Before you put up a tower, look for any overhead electrical wires. Make sure that the tower is installed where there is no possibility of contact between the lines and the tower if the guy wires should ever break or the tower fall.

T0B07 How should the guy wires for an antenna tower be installed?

A. So each guy wire anchor point has an even number of wires
B. So that no guy wire is more than 25 feet long
C. Each guy wire must be pulled as tight as possible
D. In accordance with the tower manufacturer's instructions

D The tower manufacturer will provide instructions showing the required number and placement of guys for your towers. Follow the manufacturer's instructions! They have done the calculation and testing to support their instructions. You may be required to comply with manufacturer's instructions by your building permit, as well.

T0B08 What is a safe distance from a power line to allow when installing an antenna?

A. Half the width of your property unless the wires are at least 23 feet high
B. 12.5 feet in most metropolitan areas
C. 36 meters plus ½ wavelength at the operating frequency
D. So that if the antenna falls unexpectedly, no part of it can come closer than 10 feet to the power wires

D 10 feet of separation if the tower and antenna fall directly towards the power lines is a minimum amount for safety. Figure the separation from the very top of the antenna or with the antenna oriented so that it is closest to the power lines.

T0B09 What is the most important safety rule to remember when using a crank-up tower?

A. This type of tower must never be painted
B. Crank up towers must be raised and lowered frequently to keep them properly lubricated
C. Winch cables must be specially rated for use on this type of tower
D. A crank-up tower should never be climbed unless it is in the fully lowered position

D Climbing a crank-up tower places your hands and feet between rungs and braces that can do a lot of damage if the tower sections slip or a cable breaks. Not only should you be sure the tower is fully nested, but also place a safety block such as piece of pipe or a 2-by-4 between the rungs to prevent movement.

T0B10 **Why is stainless steel hardware used on many antennas instead of other metals?**

A. Stainless steel is a better electrical conductor
B. Stainless steel weighs less than other metals
C. Stainless steel parts are much less likely to corrode
D. Stainless steel costs less than other metals

C Being exposed to the weather, corrosion is always a concern for tower and antenna builders. Most antenna manufacturers use a combination of stainless steel and aluminum because these metals resist corrosion. Tower hardware and guy lines are often galvanized with zinc which does not corrode in contact with stainless steel. Avoid the use of nickel- or cadmium-plated hardware in outdoor applications as it will quickly rust.

T0B11 **What is considered to be an adequate ground for a tower?**

A. A single 4 foot ground rod, driven into the earth no more than 12 inches from the base
B. A screen of 120 radial wires
C. Separate 8 foot long ground rods for each tower leg, bonded to the tower and each other
D. A connection between the tower base and a cold water pipe

C Don't confuse a safety ground with an RF ground! A tower's safety ground is intended to conduct any lightning energy to the earth, reducing the amount travelling along your feed lines. Grounding each leg of the tower balances lightning currents—a weak ground will encourage more energy to flow along your feed lines.

T0C RF hazards, radiation exposure, RF heating hazards, proximity to antennas, recognized safe power levels, handheld transceiver safety, exposure to others – 1 exam question

T0C01 **What type of radiation are VHF and UHF radio signals?**

A. Gamma radiation
B. Ionizing radiation
C. Alpha radiation
D. Non-ionizing radiation

D Radio and lower frequency waves are classified as non-ionizing radiation because the frequency is too low for them to ionize atoms, no matter how intense the power density of the wave may be. The frequency of **ionizing radiation** must be higher than that of visible light--ultraviolet, X-rays, gamma rays. Those types of radiation can separate electrons from atoms, creating ions.

T0C02 When can radio waves cause injury to the human body?

A. Only when the frequency is below 30 MHz
B. Only if the combination of signal strength and frequency cause excessive power to be absorbed
C. Only when the frequency is greater than 30 MHz
D. Only when transmitter power exceeds 50 watts

B Heating caused by absorption of RF by the body varies with signal strength and frequency so it is not possible to give a single safe power threshold. The FCC Rules about maximum permissible exposures to RF take into account both frequency and power.

T0C03 What is the maximum power level that an amateur radio station may use at frequencies above 30 MHz before an RF exposure evaluation is required?

A. 1500 watts PEP transmitter output
B. 1 watt forward power
C. 50 watts PEP at the antenna
D. 50 watts PEP reflected power

C [97.13(c)(1)] - At power levels below 50 watts, the FCC has determined that there is little risk to people. Stations operating with less than 50 watts above 30 MHz are categorically excluded from having to perform a station evaluation.

T0C04 What factors affect the RF exposure of people near an amateur transmitter?

A. Frequency and power level of the RF field
B. Distance from the antenna to a person
C. Radiation pattern of the antenna
D. All of these answers are correct

D The human body absorbs less RF energy at some frequencies and more at others. If you decrease your transmitter output power, you decrease the RF energy radiated from your antenna. Placing antennas farther from people reduces the power density to which they are exposed. Finally, the radiation pattern of the antenna affects where RF exposure will be greatest.

T0C05 Why must the frequency of an RF source be considered when evaluating RF radiation exposure?

A. Lower frequency RF fields have more energy than higher frequency fields
B. Lower frequency RF fields do not penetrate the human body
C. Higher frequency RF fields are transient in nature and do not affect the human body
D. The human body absorbs more RF energy at some frequencies than others

D At frequencies near the body's natural resonant frequency, RF energy is absorbed more efficiently, and maximum heating occurs. In adults, this frequency usually is about 35 MHz if the person is grounded, and about 70 MHz if the person's body is insulated from the ground. Also, body parts may be resonant as well; the adult head, for example is resonant around 400 MHz. Body size thus determines the frequency at which most RF energy is absorbed. As the frequency is increased above resonance, less RF heating generally occurs.

T0C06 How can you determine that your station complies with FCC RF exposure regulations?

A. By calculation based on FCC OET Bulletin 65
B. By calculation based on computer modeling
C. By measurement of field strength using calibrated equipment
D. All of these choices are correct

D [97.13(c)(1)] - You may use a variety of methods to determine that your station complies with FCC RF-exposure regulations. All of the choices given above are correct and valid methods of making that determination.

T0C07 What could happen if a person accidentally touched your antenna while you were transmitting?

A. Touching the antenna could cause television interference
B. They might receive a painful RF burn injury
C. They would be able to hear what you are saying
D. Nothing

B An RF burn is caused by localized heating of the body at the point of contact with the antenna. While painful, it is rarely serious. This is a good reason why you should install your antenna where people cannot accidentally come in contact with it.

T0C08 **What action might amateur operators take to prevent exposure to RF radiation in excess of FCC supplied limits?**

A. Alter antenna patterns
B. Relocate antennas
C. Change station parameters such as frequency or power
D. All of these answers are correct

D Anything you can do to reduce power density in an area of concern will reduce RF exposure. This includes relocating your antennas farther from the area, changing the antenna's radiation pattern, and reducing power. The exposure limits also change with frequency, so changing frequency to operate on a band with a higher safe exposure limit is also acceptable.

T0C09 **How can you make sure your station stays in compliance with RF safety regulations?**

A. Compliance is not necessary
B. By re-evaluating the station whenever an item of equipment is changed
C. By making sure your antennas have a low SWR
D. By installing a low pass filter

B Whenever you make a change to something that affects power density around your antennas, you should re-evaluate the station. For example, adding an amplifier or changing to an antenna that has more gain will increase power density and you should re-evaluate. If your station is already in compliance and you take a step that decreases RF exposure, such as raising your antennas farther from areas where people are, you don't need to re-evaluate.

T0C10 **Which of the following units of measurement is used to measure RF radiation exposure?**

A. Milliwatts per square centimeter
B. Megohms per square meter
C. Microfarads per foot
D. Megahertz per second

A RF exposure is measured in terms of power density (power per unit of area). Occasionally, electric or magnetic field strength (volts/meter or amps/meter) can be used in calculations, but power density in milliwatts per square centimeter (mw/cm^2) is the standard unit of power density.

T0C11 Why is duty cycle one of the factors used to determine safe RF radiation exposure levels?

A. It takes into account the amount of time the transmitter is operating
B. It takes into account the transmitter power-supply rating
C. It takes into account the antenna feed line loss
D. It takes into account the thermal effects of the final amplifier

A Because the effect of RF exposure is heating, average exposure is what is important. Average exposure during any given period depends on how long the transmitter is operating—which is measured by duty cycle. An emission with a lower duty cycle produces less RF exposure for the same PEP output.

Notes

Notes

Notes

Notes

Notes

Notes

Notes

Notes

About the ARRL

The seed for Amateur Radio was planted in the 1890s, when Guglielmo Marconi began his experiments in wireless telegraphy. Soon he was joined by dozens, then hundreds, of others who were enthusiastic about sending and receiving messages through the air—some with a commercial interest, but others solely out of a love for this new communications medium. The United States government began licensing Amateur Radio operators in 1912.

By 1914, there were thousands of Amateur Radio operators—hams—in the United States. Hiram Percy Maxim, a leading Hartford, Connecticut inventor and industrialist, saw the need for an organization to band together this fledgling group of radio experimenters. In May 1914 he founded the American Radio Relay League (ARRL) to meet that need.

Today ARRL, with approximately 150,000 members, is the largest organization of radio amateurs in the United States. The ARRL is a not-for-profit organization that:

- promotes interest in Amateur Radio communications and experimentation
- represents US radio amateurs in legislative matters, and
- maintains fraternalism and a high standard of conduct among Amateur Radio operators.

At ARRL headquarters in the Hartford suburb of Newington, the staff helps serve the needs of members. ARRL is also International Secretariat for the International Amateur Radio Union, which is made up of similar societies in 150 countries around the world.

ARRL publishes the monthly journal *QST*, as well as newsletters and many publications covering all aspects of Amateur Radio. Its headquarters station, W1AW, transmits bulletins of interest to radio amateurs and Morse code practice sessions. The ARRL also coordinates an extensive field organization, which includes volunteers who provide technical information and other support services for radio amateurs as well as communications for public-service activities. In addition, ARRL represents US amateurs with the Federal Communications Commission and other government agencies in the US and abroad.

Membership in ARRL means much more than receiving *QST* each month. In addition to the services already described, ARRL offers membership services on a personal level, such as the ARRL Volunteer Examiner Coordinator Program and a QSL bureau.

Full ARRL membership (available only to licensed radio amateurs) gives you a voice in how the affairs of the organization are governed. ARRL policy is set by a Board of Directors (one from each of 15 Divisions). Each year, one-third of the ARRL Board of Directors stands for election by the full members they represent. The day-to-day operation of ARRL HQ is managed by an Executive Vice President and his staff.

No matter what aspect of Amateur Radio attracts you, ARRL membership is relevant and important. There would be no Amateur Radio as we know it today were it not for the ARRL. We would be happy to welcome you as a member! (An Amateur Radio license is not required for Associate Membership.) For more information about ARRL and answers to any questions you may have about Amateur Radio, write or call:

ARRL—The national association for Amateur Radio
225 Main Street
Newington CT 06111-1494

Voice: 860-594-0200
Fax: 860-594-0259

E-mail: **hq@arrl.org**
Internet: **www.arrl.org/**

Prospective new amateurs call (toll-free):
800-32-NEW HAM (800-326-3942)
You can also contact us via e-mail at **newham@arrl.org**
or check out **ARRLWeb** at **www.arrl.org**

FEEDBACK

Please use this form to give us your comments on this book and what you'd like to see in future editions, or e-mail us at **pubsfdbk@arrl.org** (publications feedback). If you use e-mail, please include your name, call, e-mail address and the book title, edition and printing in the body of your message. Also indicate whether or not you are an ARRL member.

Please check the box that best answers these questions:

How well did this book prepare you for your exam?
☐ Very Well ☐ Fairly Well ☐ Not Very Well

Which exam did you take (or will you be taking)?
☐ Technician ☐ Technician with code ☐ General

Did you pass? ☐ Yes ☐ No

Do you expect to learn Morse code some time? ☐ Yes ☐ No ☐ Already know code

Where did you purchase this book?
☐ From ARRL directly ☐ From an ARRL dealer

Is there a dealer who carries ARRL publications within:
☐ 5 miles ☐ 15 miles ☐ 30 miles of your location? ☐ Not sure.

If licensed, what is your license class? ______________________________

Name ______________________________ ARRL member? ☐ Yes ☐ No

Call Sign ______________

Address ______________________________

City, State/Province, ZIP/Postal Code ______________________________

Daytime Phone (   ) ______________ Age ________ E-mail ______________

If licensed, how long? ______________________________

Other hobbies ______________________________

____

Occupation ______________________________

For ARRL use only	T Q&A
Edition	4 5 6 7 8 9 10 11
Printing	1 2 3 4 5 6 7 8 9 10 11

From ______________________________

____

____

Please affix postage. Post Office will not deliver without postage.

EDITOR, ARRL'S TECH Q&A
ARRL—THE NATIONAL ASSOCIATION FOR AMATEUR RADIO
225 MAIN STREET
NEWINGTON CT 06111-1494

....please fold and tape...